Quaternary Ecology

Quaternary Ecology
A paleoecological perspective

Hazel R. Delcourt and Paul A. Delcourt
University of Tennessee, USA

CHAPMAN & HALL
London · New York · Tokyo · Melbourne · Madras

UK   Chapman & Hall, 2–6 Boundary Row, London SE1 8HN
USA   Chapman & Hall, 29 West 35th Street, New York NY10001
JAPAN   Chapman & Hall Japan, Thomson Publishing Japan, Hirakawacho
            Nemoto Building, 7F, 1–7–11 Hirakawa-cho, Chiyoda-ku,
            Tokyo 102
AUSTRALIA   Chapman & Hall Australia, Thomas Nelson Australia,
            102 Dodds Street, South Melbourne, Victoria 3205
INDIA   Chapman & Hall India, R. Seshadri, 32 Second Main Road,
            CIT East, Madras 600 035

First edition 1991

© 1991 Hazel R. Delcourt and Paul A. Delcourt

Typeset in 10/12pt Plantin by Excel Typesetters Company,
Hong Kong
Printed in Great Britain by T. J. Press (Padstow) Ltd, Padstow,
Cornwall

ISBN 0 412 29780 9 (HB)
     0 412 29790 6 (PB)

British Library Cataloguing in Publication Data
Delcourt, Hazel R.
    Quaternary ecology.
    I. Title II.   Delcourt, Paul A.
    560

ISBN 0–412–29780–9
ISBN 0–412–29790–6 pbk

Library of Congress Cataloging-in-Publication Data
Available

This book is dedicated to
Michelle Jeannette Delcourt
whose life makes all the rest worthwhile

Contents

Preface

Over the past several decades the field of Quaternary ecology has undergone a fundamental change in perspective. Paleoecologists, who were formerly concerned with biostratigraphic questions, have increasingly begun to develop and test explicitly ecological hypotheses. Literature emphasizing the contributions of paleoecology to contemporary issues in ecology is growing, but many of the key papers are published in journals not traditionally read by ecologists. The tendency toward increased specialization within ecology, along with the proliferation of new journals as publication outlets, increasingly makes it difficult to communicate effectively across subdisciplines within ecology. With this book, we hope to bridge the communication gap between Quaternary ecologists and other ecologists.

In this book we do not attempt to cover the subject of Quaternary ecology in a traditional textbook presentation. Two comprehensive books, one contemporary text (Birks and Birks, 1980) and a handbook of techniques (Berglund, 1986) appropriate to various specialties within this field are available to the reader who is interested in the details of methods used in reconstructing past communities and ecosystems.

This volume is intended to supplement traditional textbooks in ecology and Quaternary ecology by examining important issues and controversies in ecology that can be approached fruitfully using paleoecological methods. We use examples from the Quaternary literature to show how this expanded viewpoint of ecology can be of relevance to a broad audience of ecologists and to highlight its potential for contributing new insights on important issues. In our choice of examples we guide the reader to literature sources that contain a wealth of additional references to papers on topics of more specialized interest. For example, the following are continental-scale syntheses of the Quaternary history of terrestrial ecosystems: Delcourt and Delcourt (1987a); Graham *et al.* (1987); Huntley and Birks (1983); Huntley and Webb (1988); Martin and Klein (1984); Porter (1983); Ritchie (1987); Ruddiman and Wright (1987); Velichko *et al.* (1984); and Wright (1983). Recent syntheses of the Quaternary development of freshwater aquatic ecosystems (paleolimnology) include: Haworth and Lund (1984); Merilainen *et al.* (1983); and Smol *et al.* (1986). Major paleo-oceanographic works include: Cline and Hays (1976); CLIMAP (1981); and Ruddiman and Wright (1987).

In Chapter 1 we provide an introduction to the fundamental principles and underlying assumptions of paleoecology. These are general principles applicable over a range of time scales. We also consider the importance of an expanded view of spatial and temporal scale as fundamental for placing modern ecosystems in context. For the remainder of the book we focus specifically upon a series of ecological issues that are resolvable on the time scale most close to the experience of ecologists, the Quaternary Period of geological history (the past several million years). This is the span of time during which the modern distributional ranges of the biota have taken shape and over which the modern biotic communities have assembled. Within this expansive time frame, events of the past 100 000 years, the most recent glacial–interglacial cycle, are most directly relevant to understanding the present-day structure and functioning of terrestrial and aquatic communities and ecosystems. Most of our examples deal with 'natural experiments of the past' drawn from the time intervals of the late Pleistocene (from 100 000 to 10 000 years ago) and the Holocene interglacial (the past 10 000 years). In the last chapter we consider how information about the biotic communities of the past can be applied to questions of conservation of biological diversity and future global change.

We would like to acknowledge Dr Alan Crowden for his encouragement in suggesting that we write this book, as well as Dr H.J.B. Birks and Dr Stuart Pimm for their thoughtful and constructive reviews. We thank the Department of Geological Sciences and the Graduate Program in Ecology of the University of Tennessee, Knoxville, for providing us both with a research semester in the fall of 1989, during which the first draft of this manuscript was completed. This is Contribution No. 53 from the Center for Quaternary Studies of the Southeastern United States, University of Tennessee, Knoxville.

1 The paleoecological perspective

Paleoecology is the study of individuals, populations and communities of plants and animals that lived in the past and their interactions with and dynamic responses to changing environments. Modern ecosystems represent only a portion of the ecosystems that have existed over the long time scale of the Earth's history. The paleoecological perspective is thus an expanded view of ecology that takes into account changes through relatively long intervals of time. By studying former plant and animal communities and their changes over time, it is possible to understand more fully the history of development, structure and function of modern ecosystems.

1.1 ECOLOGICAL AND EVOLUTIONARY TIME SCALES

The flora and fauna of the oceans and the terrestrial biosphere have evolved over billions of years. Both marine and terrestrial ecosystems have functioned throughout geological history, but their structure and composition have changed through time as global conditions of atmosphere, oceans, and continents have changed.

Processes with which paleobiologists are concerned in evolutionary time include changes in the rates of extinction and of speciation. Through geological history, periods of relative equilibrium have been punctuated by major changes in extinction and speciation rates (Stanley, 1979). These macroevolutionary events may be caused by changes in physical boundary conditions such as plate-tectonic configurations of continents and seas that determine global geography. Environmental changes such as the development of an oxygen-rich atmosphere, which occurred two billion years ago (Cloud, 1976), have resulted in fundamental new conditions that have had profound effects on the nature of the biota. For example, the increase of oxygen within both Precambrian oceans and atmosphere triggered the ecological displacement of anaerobic forms of marine life by oxygen-tolerant organisms, a prerequisite for the much later morphological and physiological adaptations that enabled colonization of land by plants and animals in the Silurian Period, 420 million years ago (Gensel and Andrews, 1987).

In ecological time, microevolutionary processes occur, such as character displacement in populations resulting from predation pressure. The field of 'evolutionary ecology' is primarily concerned with questions of population genetics and adaptation on a relatively short time scale compared with that dealt with by paleoecologists (Emlen, 1973).

Ecological time is thus embedded within evolutionary time, and to understand ecological patterns and processes at any given time it is important to analyze the distribution and abundance of individuals of the species as well as their interrelationships with each other and their relationship to their physical environment (Valentine, 1973; Nitecki, 1981).

1.1.1 Uniformitarianism

'The present is the key to understanding the past' is a fundamental starting point for reconstructing ecosystem changes through long periods of time. The principle of uniformitarianism states that physical, chemical, and biological processes operative today are ones that have governed throughout the Earth's history; therefore, given enough evidence in the form of fossils, and in sedimentological, stratigraphical and biogeochemical data, it is possible to reconstruct and interpret past environments and biotic communities based upon natural processes operative today (Gould, 1965; Valentine, 1973).

The intensity and rate at which physical processes occur, however, vary through time. For example, glaciers occur today, including the Greenland ice cap, the Antarctic ice cap, and mountain glaciers at high latitudes or high elevations located, for instance, in the northern Rocky Mountains of Alaska, the Alps, the Andes and the Himalayas. At the height of the last Ice Age, 18 000 years ago, glaciers covered a much larger area, particularly over the Northern Hemisphere, and were a much more important influence on the dynamics of global climate and biota than they are today. Through the last several million years of the Quaternary geological period, the rate and intensity of climatic fluctuations has changed, with the last seven glacial–interglacial cycles becoming 'saw-toothed' in nature (Fig. 1.1); that is, during each glacial–interglacial cycle of 100 000 years, 90 000 years of gradual global climatic cooling is followed by rapid warming and 10 000 years of interglacial warmth (Imbrie and Imbrie, 1979; Pisias and Moore, 1981).

The challenge for the paleoecologist is to reconstruct in as much detail as possible the former patterns of distribution and abundances of individuals and populations, their physiology and functional relationships based on their morphology and biochemistry, and the paleogeography of the environments in which they lived; then to infer the biotic and abiotic processes that resulted in the observed paleo-ecosystem structure and that constituted the functioning of ecosystems in the past. A uniformitarianistic approach does not necessarily mean that past biological communities have strict modern analogues in terms of their taxonomic composition, trophic levels repre-

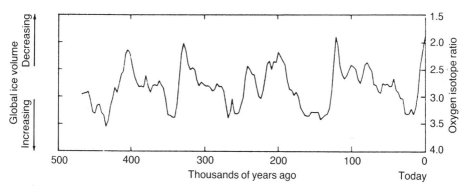

Figure 1.1 Record of climatic changes over the past 500 000 years as measured by oxygen isotope ratios from cores of deep-sea sediments obtained from the Indian Ocean. From Imbrie and Imbrie (1979).

sented, or specific niches occupied. As the atmosphere, oceans, continents and global climate have changed through time, the degree of analogue in composition and structure of past communities has varied from complete dissimilarity to very close analogy when compared with today's communities.

1.1.2 Analogue concept

Reasoning by analogy is fundamental to paleoecology. By determining how closely analogous past biotic communities are to modern ones we gain a measure of absolute amount and rate of ecological change through time. The differences in community composition can be measured quantitatively using dissimilarity coefficients (Prentice, 1986a). The direction and rate of community change through time can be measured by plotting dissimilarity values for a series of paleoecological sites on maps at different times. In theory, there are five possible relationships of past communities to modern communities that can be illustrated with a series of Venn diagrams (Fig. 1.2). Communities of the past may completely overlap in composition with each other or with modern communities (Fig. 1.2(a)), in which case the calculated amount of dissimilarity would approach zero and the past communities would be said to have excellent modern analogues. Composition of past and present communities may only partially overlap (Fig. 1.2(b)), either taxonomically or in terms of percentages of taxa comprising the community (Hutson, 1977). In this case, the past community would be categorized as either having relatively good or relatively poor modern analogues, for most dissimilarity coefficients depending on the numerical cut-off value chosen between zero (identical in composition) and one (completely dissimilar) (Prentice, 1986a). One way to determine whether a modern analogue is good or is poor is by calibrating the dissimilarity coefficients using the composi-

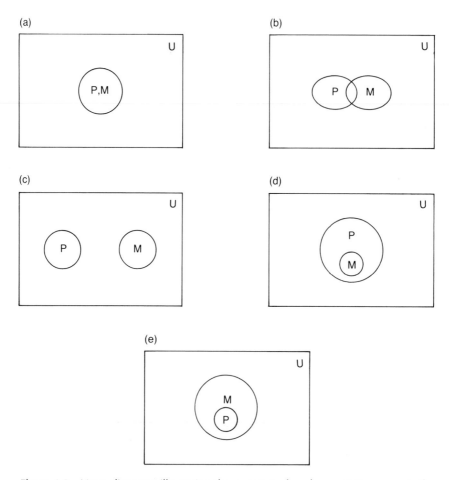

Figure 1.2   Venn diagrams illustrating the concept of analogue; 'U' represents the universe of all possible communities, 'P' represents the subset of past communities, and 'M' represents the subset of modern communities: (a) past communities have an excellent modern analogue; (b) past communities have a partial modern analogue; (c) past communities have no modern analogue; (d) modern communities are a subset of all past communities; (e) past communities are a subset of all modern communities.

tional differences between modern communities arrayed along climatic, physiographical, or other environmental gradients (Overpeck *et al.*, 1985). If paleo-communities are demonstrated to be completely different from modern communities (Fig. 1.2(c)), they are said to have no modern analogues. The modern community may represent only a subset of the former communities (Fig. 1.2(d)), in which case it will have a range of dissimilarity values when compared with past community compositions from different time ranges and across a variety of geographical settings. The fifth possibility (Fig. 1.2(e)) is that all past communities may be included as a subset within the suite of all modern communities; however, the fossil record indicates that this condition has not been true during the history of the Earth (Cloud, 1976).

Before analyzing the fossil record to determine which of the five Venn diagram scenarios is applicable, it is first necessary to define what is meant by 'modern analogues'. In eastern North America, for example, plant ecologists have used three different kinds of definitions of modern communities. A literal interpretation includes all communities that exist today and that can be sampled directly, regardless of whether they have been impacted by human activities or contain non-native species (Burgess and Sharpe, 1981; Delcourt *et al.*, 1984). Most North American ecologists have instead been interested in documenting the composition of natural vegetation as it existed in presettlement times, that is, prior to disturbance by EuroAmerican settlers, by sampling old-growth or 'virgin' stands of vegetation or by using early land surveys that recorded 'witness trees' on a systematic grid prior to settlement of the land (Braun, 1950; Grimm, 1984). An even broader approach is used by some Quaternary ecologists who consider that the late-Holocene time of 500 years ago represents the pre-Columbian time-line for natural vegetation, and who therefore must rely primarily on the plant-fossil record to obtain a measure of the vegetation composition used as the 'modern analogue' for comparison with the composition of previous vegetation (Jacobson *et al.*, 1987).

An example of the application of the analogue approach in Quaternary ecology is the study by Delcourt and Delcourt (1987a, b) of the compositional changes in forests of eastern North America over the past 20 000 years. Using fossil pollen data from 162 radiocarbon-dated lake sites distributed across the region, and calibrations of pollen representation of 19 major forest taxa based upon the correspondence of modern pollen assemblages with measures of tree dominance in commercial forests, past composition of forest communities was quantified for six times since the last full-glacial interval (Fig. 1.3). The composition of the forest community represented by each fossil sample was compared with the composition of all other samples to determine the degree of similarity among samples from different times and across a large geographic area. Samples were arrayed in ecological space using the standard ecological method of ordination called Detrended Correspondence Analysis (DCA) (see Gauch, 1982, for a discussion of this numerical technique). This technique calculates the compositional dissimilarity between samples and plots them in ordination space using an x–y plot. The arrangement of samples along the x-axis (Axis 1 in Fig. 1.3) accounts for the greatest amount of variability in the data set; the y-axis (Axis 2) accounts for the second greatest source of variability. Samples located together in tight clusters in DCA ordination space are very similar in

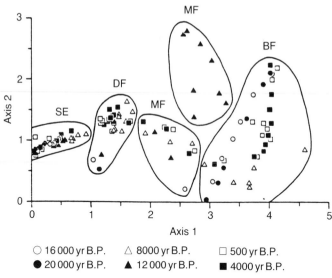

Figure 1.3 Forest communities of eastern North America for the past 20 000 years plotted on the first two ordination axes determined by Detrended Correspondence Analysis (DCA). Symbol codes designate the age of the samples of reconstructed forest compostition. The presettlement forest communities for 500 years ago indicated by the open box symbol provide the suite of modern communities (prior to disturbance by EuroAmericans) used as possible modern analogues to compare with much older forest communities. Clusters denote the ordination space represented by the southeastern evergreen forest (SE), deciduous forest (DF), mixed conifer-northern hardwoods forest (MF, lower cluster with modern analogues, upper cluster without modern analogues), and boreal forest (BF). Modified from Delcourt and Delcourt (1987a,b).

composition to each other, whereas samples located farther apart are less similar in composition to each other. In Fig. 1.3 the forest communities for 500 years Before Present (yr B.P.) represent presettlement forests. That is, the presettlement forest samples represent essentially modern communities as they existed prior to recent disturbance by EuroAmericans. The samples for 500 yr B.P. were used in this study as possible modern analogues to compare with much older forest communities. This techique therefore provided a quantitative means of identifying forest communities that have persisted throughout the last glacial–interglacial cycle as well as those that have been ephemeral and for which only poor modern analogues exist.

Several distinct clusters of samples were identified by DCA results (Fig. 1.3). The first axis of variation on the ordination plot (Fig. 1.3) represents the latitudinal gradient, from warm–temperate to boreal forests, and accounts for most of the compositional variation within the paleovegetational data set. Samples representing the southeastern evergreen forest (SE in Fig.

1.3) clustered tightly together in ordination space, indicating similarity in composition (and hence relatively good modern analogues) throughout the last 20 000 years. The southeastern evergreen forest has maintained compositional stability and has remained within the same geographical area throughout this time interval. In contrast, the broad cluster of samples for the boreal forest (BF in Fig. 1.3) contains samples from full-glacial times (20 000 yr B.P. and 16 000 yr B.P.) as well as samples from full-interglacial times (8000 yr B.P. to the present). The full-glacial samples of boreal forest have good modern analogues; they are comparable both in the assemblage of tree taxa and in the range of compositional heterogeneity. Although Pleistocene boreal forests have good modern analogues in terms of composition, the geographic location of the southern limit of boreal forest has shifted more than 1000 kilometers during the last 20 000 years (Delcourt and Delcourt, 1987a). The deciduous forest cluster (DF in Fig. 1.3) contains samples only from 8000 years ago to the present, indicating that it has emerged as a new compositional type during the present interglacial period. Samples from the mixed conifer–northern hardwoods forest (MF in Fig. 1.3) plot within two different clusters in DCA ordination space. The lower cluster contains good modern analogues for the past 16 000 years. The upper cluster represents samples from 12 000 years ago that are compositionally similar to each other but that lack good modern analogues (they are dissimilar to any forest communities that existed in eastern North America 500 years ago). Hence, these late-glacial forests were ephemeral communities that can be explained by major climatic warming that resulted in unstable landscapes and differential species migrations (Delcourt and Delcourt, 1983, 1987a, b).

This study documents that Fig. 1.2(b) is the most realistic Venn diagram representation for the Quaternary Period. Over the last 20 000 years some past communities have existed which have excellent modern analogues. However, there have also been some past combinations of species for which there is no counterpart in modern communities, and there are certain modern communities that have assembled only recently and that have no past precedent.

1.2   RESEARCH DESIGN

Development of statistically valid experimental design is central to contemporary research in ecology. Paleoecology differs in its methods from some other branches of ecology in that investigators are not able to establish an experiment and replicate it (Hurlbert, 1984). Rather, paleoecologists look for evidence of events that occurred through time and use these 'natural experiments of the past' to test null hypotheses (Deevey, 1969; Diamond, 1986). By analyzing time-series of empirical data based on the history of events in ecological and evolutionary time, paleoecologists can directly test theoretical models that predict responses of ecological systems to disturbances or long-term environmental changes (Shugart, 1984; Prentice, 1986a).

In paleoecology, as in any field of scientific endeavor, the development of an appropriate research design is guided by the specific research question asked, the ecological hypothesis to be tested. The initial statement of the problem influences the selection of the level of biological organization to be studied (individual, population, community or ecosystem). Important decisions at the outset include the type of organism or even the specific group of organisms, the nature of the environment (terrestrial, aquatic or aerial) and the prevalent disturbance regime (the kind, intensity and recurrence interval of disturbances), the spatial bounds established for the study area, and the temporal range or duration of the phenomenon to be investigated. Additional decisions involve the type or suite of most suitable techniques for sampling the characteristics and interactions among organisms and with their environment, statistical methods and adequate sample size, as well as selection of sampling interval to ensure appropriate resolution in spatial, temporal and taxonomic data collected.

Ecologists can potentially study all aspects of ecosystems across all spatial scales represented on the surface of the Earth today. However, although it may be technically possible for contemporary scientists to focus their research on specific population dynamics, nutrient cycling or energy transfer across a landscape mosaic, no modern ecosystem yet has been studied completely. Few ecosystems have been comprehensively monitored for more than the last several decades. In North America the empirical data sets from even the oldest established Long-Term Ecological Research sites such as the Hubbard Brook Experimental Watersheds in New Hampshire (Likens *et al.*, 1977; Bormann and Likens, 1979; Likens, 1985) and Coweeta Experimental Watershed in western North Carolina (Swank and Crossley, 1988) offer insights into the long-term development of natural ecosystems that can be augmented substantially with paleoecological data representing an expanded time interval (M. Davis, 1989a).

Traditionally, ecologists have studied plant succession and recovery of ecosystems after disturbance by substituting space for time (Pickett, 1989). In doing so, it is assumed that the spatial heterogeneity of successional stages of modern communities distributed across the landscape will accurately reflect the sequence of changes in a system following an episode of disturbance or occurring during primary succession. If this assumption does not hold true for the system studied, it is not possible to characterize in real time the dynamic changes in natural populations and communities as they respond to both internal and external forces (Pickett, 1989). Rather, preconceived ideas of ecosystem change will condition and limit the general applicability of the results from the study. This research objective does not pose an intractible dilemma; rather, this logistical problem in procedure can be solved by adding new, complementary tools of investigation. By collecting the appropriate kinds of data, paleoecologists can obtain both the spatial and temporal resolution needed in order to test the validity of assumptions such as that of substitution of space for time in successional studies. In this way, a paleoecological approach can both open a vista into the past and can allow us successfully to examine long-term ecosystem dynamics.

1.3 PRINCIPLES OF PALEOECOLOGICAL RECONSTRUCTION

Paleoecologists are constrained in their ability to recover meaningful paleoecological information by several attributes of the systems they study. The fossil assemblage may be completely preserved, or the assemblage may be distorted by geological and biological processes that act upon it during and after it is deposited within a sedimentary environment. In order to draw appropriate conclusions from paleoecological data, the investigator must examine data sets for evidence of distortion caused by events occurring between the time the assemblage was deposited and the time it was recovered for study. In order to avoid misinterpretation of the fossil evidence, this type of investigation is necessary before we can begin meaningful ecological analysis of the data (Birks and Birks, 1980).

1.3.1 Preservation of organic remains as fossils

Fossil evidence of formerly living organisms may be either the remains of the organisms or indirect evidence of their behavior, consisting of traces, impressions, or burrows in sediments, or of borings on rock surfaces. Only those organisms that have been preserved in some way in the fossil record can be used in paleoecological reconstructions. Organisms with bodies composed entirely of soft organic tissue are highly vulnerable to decomposition after death, unless they are buried in place immediately by sediments or moved quickly to a nearby aquatic environment and buried there by sediments. Organic remains of soft-bodied organisms, including bacteria and cyano-bacteria (blue-green algae) are the oldest known fossil forms of life, preserved in marine deposits dating from 3.8 billion years ago. Trace fossils of soft-bodied, multicellular animals have been discovered as flattened 'body impressions' in marine deposits back to about 700 million years ago (Glaessner, 1984). In contrast, organisms that produce dense mineralized or chemically resistant body parts (such as skeletons or shells) are less likely to be completely decomposed after death. Animal bones and shells, primarily constructed from minerals such as quartz, calcite, aragonite or apatite (as in the case of human bone), are found in the fossil record during the last 570 million years (Glaessner, 1984). The outermost organic wall of both pollen grains and spores is composed of a form of cellulose that is extremely resistant to biochemical attack or chemical reaction. These microfossils of terrestrial plants have persisted in sedimentary rocks for many hundreds of millions of years (Gensel and Andrews, 1987). Opaline phytoliths, quartz nodules secreted within living cells of vascular plants, may survive intact within soils for thousands of years.

Organisms are most likely to be preserved as fossils if, upon death, they are deposited and buried rapidly within an environment that favors preservation of organic remains. Burial by fine-grained silt and clay particles can provide a protective seal that will inhibit further decomposition by *in situ* microbes. Environmental conditions that limit microbial attack include constant moisture (either permanently wet or permanently dry), low pH as found in *Sphagnum* bogs, oxygen-deficient sediments such as muds at the bottom of a deep lake, and relatively cold temperatures or even permafrost (Sangster and Dale, 1964; Kurten, 1986). Different kinds of sites where sediments accumulate have different potentials for preserving organic remains. In tropical and temperate regions, fossil assemblages rarely persist on upland surfaces that are subject to erosion or to seasonal wet–dry or cold–warm cycles. In boreal or arctic regions, a more constant environment for preservation is provided by perennially frozen ground. Optimal environments for preserving fossil remains include freshwater lakes, bogs, swamps, brackish-water estuaries, and ocean basins. Therefore, organisms living in or near aquatic habitats are more consistently represented in the fossil record than are those from terrestrial habitats. This is especially the case for macroscopic remains that are not easily transported by wind or water from uplands distant from the wetland environment. The most important exception to this rule is the fossil record of pollen grains and spores, which typically contains consistent representation of plant taxa from both terrestrial and aquatic ecosystems (Birks and Birks, 1980).

In certain circumstances, observed fossil assemblages may represent an accurate sample of former communities of organisms that died and were buried by sediments at the site where the plants and animals initially lived. In peat bogs, strongly acidic water at the bog surface both allows for rapid vertical growth of *Sphagnum* mosses and ensures the *in situ* preservation of many of the plants (Barber, 1982). In other circumstances, however, fossil assemblages may not reflect the initial community composition in its entirety because of biological and geological processes occurring before, during, and after the organic remains are deposited. For example, the thick cuticle of an evergreen magnolia (*Magnolia*) leaf may allow it to be more resistant to decay than the more delicate leaf of a deciduous maple (*Acer*). Thus, when a leaf mat which has accumulated on a swamp forest floor is exposed to biological decomposition, the remains of certain plant species within the assemblage may be destroyed preferentially. In this case, the resulting fossilized residue contains less taxonomic and ecological information than the original assemblage. The selective loss of plant taxa through such deterioration after deposition results in information gaps that affect the completeness with which the original community can be reconstructed (Delcourt and Delcourt, 1980). Indices of preservation quality provide measurable criteria by which to recognize and to quantify the magnitude of information lost through differential degradation of the fossil assemblage. Such indices give insight into the completeness of paleoecological reconstructions with which tests of ecological hypotheses will be made (Delcourt and Delcourt, 1980; Hall, 1981).

The original sample of the former community can be distorted further by hydrological processes. The initial assemblage may be sorted differentially as water currents physically winnow the organic remains on the basis of their size, shape and density. Removal by sorting distorts the composition of the fossil assemblage and eliminates information that otherwise could be recovered from the sample. Alternatively, some organic remains may be transported to another sedimentary environment where they may be added to another assemblage. This results in artificial mixing of taxa from different communities living contemporaneously within one landscape. A serious problem arises if fossils from an older geological deposit are eroded, transported, and redeposited with remnants of a geologically younger community. This produces a fossil assemblage with specimens of different ages, potentially giving a false representation of the former community and precluding evaluation of long-term community dynamics (Cushing, 1964).

1.3.2 Sampling strategies

Paleoecologists may collect assemblages of fossils from sediments (or cemented sedimentary rocks) exposed as vertical sections in cliffs or cut-banks of streams. Coring devices are used to recover long sequences of sediment cores from beneath lakes and ponds, bogs, swamps or oceans. In either situation, the objective is to collect vertical sequences of sediment that have accumulated through time in either modern or ancient environments. In a typical, undeformed sediment core from a natural lake, the oldest assemblages are at the base of the core and relatively younger assemblages occur at progressively higher positions within the sediment sequence. Modern organisms contribute to sediments forming today at the mud–water interface on the lake bottom (the top of the core sequence). The geological sequence may be continuous through time, or it may be discontinuous with time gaps in the sedimentary record. Such gaps of time can result either if sediments did not accumulate during a specific time interval or if the deposits were removed by erosion.

In addition to establishing the relative series of changes in past communities for each collection site, absolute age determinations are required to place the fossil assemblages in a context of true temporal changes. Radiometric age determinations can be made using measurements of the radioactive disintegration of unstable isotopes of elements. Each isotopic clock has a different half-life of radioactive decay and therefore an optimal span of absolute time for which it is most appropriate (see Mahaney, 1984 for a detailed discussion of Quaternary dating techniques). For example, isotopes of ^{14}C and ^{12}C can be used to date organic-rich deposits that have accumulated within the last 80 000 years. By obtaining absolute ages for sediments from a number of depths within one sediment-core sequence, gaps in time can be identified. In continuously deposited portions of the sequence, sediment accumulation rates can be estimated from the empirically determined relationships of age versus sediment core depth. Given absolute ages for each level within the sediment sequence, paleoecologists can adjust their sampling interval to analyze fossil assemblages at a temporal resolution determined by the hypothesis to be tested.

Natural lake systems are preferred for most Quaternary paleoecological investigations because many elements of research design can be controlled, and ecological phenomena can be studied over all time spans from years to millions of years. Sites can be chosen to optimize both temporal and spatial resolution of long-term dynamics of both aquatic and terrestrial ecosystems. In certain small, deep lakes of temperate regions, water circulation is limited to the upper portion of the water column and the lowermost lake water is relatively cold and dense, with little or no free dissolved oxygen. These conditions are inhospitable to most forms of aquatic bottom-dwelling life, allowing fine-grained sediments to accumulate in alternating mineral-rich and organic-rich seasonal layers. These annually laminated sediments (varves) are preserved as discrete layers in the absence of benthic organisms that otherwise might mix the horizontal layers of lake sediment. Such varved sediment sequences may span as much as the last 10 000 years (for example, Lake of the Clouds in northern Minnesota, USA; Craig, 1972).

In large or shallow lakes, both physical and biological factors may resuspend sediment once again within the water column or mix it on the lake bottom. The net result is the vertical reworking of lake muds causing mixing of materials of slightly different ages and reducing the potential temporal resolution of each sample. In most lakes of temperate and boreal regions, seasonal changes in temperature trigger two episodes of complete lake circulation, with one overturn as the water cools in fall and one in spring as lake ice melts. During overturn, water currents erode the uppermost few centimeters of lake-bottom mud, resuspending and then redepositing it. Each seasonal reworking of mud preferentially resuspends sediment from the shallow sides of the lake and transports it into the deeper central basin. At Frains Lake in south-central Michigan, USA, where atmospheric transport of pollen grains to the lake and their limnological reworking have been studied in great detail (M. Davis *et al.* 1971; M. Davis, 1973), the amount of sediment resuspended each year is equivalent to about the last two to four years net accumulation. This amount of mixing of sediment reduces the apparent annual variability of pollen influx to the lake and naturally smoothes the fossil-pollen record as a several-year running average.

In productive lakes, which are characterized by both substantial inputs of nutrients and oxygen-rich waters, the activities of bottom-dwelling organisms may mix the uppermost lake sediments over a vertical zone of many centimeters. In lakes in Maine, USA, bottom-dwelling animals live in the uppermost 4 cm but burrow down to 8 cm below the sediment surface (R. Davis, 1967). In these lakes, based upon the sediment accumulation rates of 2 mm each year, the fossil-pollen assemblages are mixed for 20 to 40 years before they are buried below the zone of biological mixing of sediments. The minimum temporal resolution that can be obtained from sampling such sites is determined by the extent of sediment mixing. Other criteria in selection of lakes with respect to temporal resolution in sampling involve choices based on lake-basin morphology, levels of nutrient input and productivity status, the limnological conditions that govern water circulation and overturn, and the characteristics of soil erosion within the watershed resulting in the rates of sediment accumulation in the lake basin.

1.3.3 Spatial resolution of fossil assemblages from Quaternary lake sediments

The spatial resolution inherent within each fossil assemblage is determined by the original source area from which the organisms are derived, their mechanisms and distances of transport, and the depositional environment within which they are entombed. Each depositional environment and each type of fossil may reflect different although diagnostic situations of preservation. As a result, special collection techniques, analytical methods, and conceptual models have been developed to focus upon the unique questions paleoecologists ask in interpreting both the sedimentary environment and geomorphological context of each site (Goudie, 1981), as well as the paleoecological significance of each different group of organisms represented in the fossil record (Grayson, 1984; Birks and Birks, 1980; Berglund, 1986). Paleoecologists commonly study closed lake basins, which do not have inflowing or outflowing streams.

Jacobson and Bradshaw (1981) have integrated models of both the mechanisms of transport (Tauber, 1965) and the nature of source area (Janssen, 1966) in order to produce the contemporary paradigm for the paleoecological interpretation of fossil assemblages of pollen grains and spores. For example, within a temperate, forested landscape the dominant forest trees produce pollen grains that may fall directly to the ground surface under the influence of gravity (the gravity component or C_g, Fig. 1.4(a)). If a wind current within the forest should capture the pollen grain, it may be carried through the open trunk space (the trunk-space component or C_t) between the forest canopy and ground surface, it may be swept over the forest immediately above the canopy (the canopy component or C_c), or it may be rafted into the atmosphere by convection and carried great distances horizontally. This last component

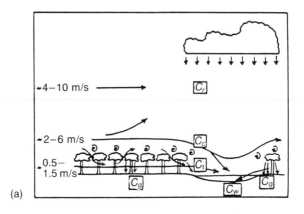

(a)

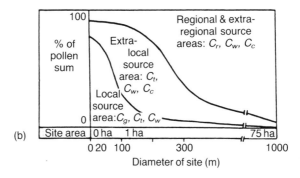

(b)

Figure 1.4 (a) Tauber's model of pollen transport within a forested landscape. Five components of pollen transport include gravity (C_g), trunk space (C_t), canopy (C_c), rain (C_r), and flowing water (C_w). (b) Jacobson and Bradshaw's model for the relationship between the diameter and area of a paleoecological site and its pollen source area as reflected by various components of the pollen assemblage. From Delcourt and Delcourt (1987a).

of pollen dispersal is designated as the rainout component (C_r) as raindrops from storms condense around the pollen grains and together the pollen and precipitation fall back to the land surface. Once on the ground surface, pollen grains may be eroded and transported by water in sheetwash or in streams and rivers (the component of water transport or C_w). These five mechanisms of pollen dispersal account for the transport of pollen assemblages to lake sites. The diameter of the lake basin determines the relative contribution of pollen grains from plants in four source areas (Janssen, 1966): local (within a 20-meter radius of the lake shore, Fig. 1.4(b)); extra-local (within the concentric zone of 20 meters to 2 kilometer-radius); regional (a distance of 2 kilometers to 200 kilometers); and extra-regional (beyond a 200-kilometer radius) (Jacobson and Bradshaw, 1981). The pollen assemblages accumulating in the smallest woodland pools (less than 0.5 hectares) are dominated by local plants growing within a 20-meter radius of the site (the pollen dispersal modes primarily by the gravity and trunk-space components). Increasingly larger pond and lake basins collect pollen assemblages produced by vegetation covering progressively larger source areas (Fig. 1.4(b)). Theoretical studies by Prentice (1985) indicate that there are substantial differences in depositional velocities for pollen types with different morphologies and densities; as a consequence, smaller, lighter pollen grains should be better represented with increasing size of lake basin. Thus, the paleoecologist can select the appropriate kind and size of lake site available in the study region in order to sample at the desired level of spatial resolution. In a given study area, the paleoecologist may be limited in the specific research questions addressed by the kinds and sizes of sites that are available as well as the time interval the sites record.

Although paleoecologists must confront a series of special biases inherent in the fossil record, their research is generally no more constrained by site factors and logistics of sample design than that of any other ecologist. In many instances, the limitations inherent in the fossil record are outweighed by its advantages. For example, on the modern landscape, it is becoming increasingly difficult to locate and study in isolation remnants of natural ecosystems that have not been disturbed by modern, post-industrial human activities. The recent history of land use is a pervasive influence upon today's biotic communties, but it is a constraint from which the paleoecologist is free in studies of prehistoric sediments. Experimental studies of modern communities must be designed using species with short-lived individuals rather than studying population dynamics of long-lived species such as trees. Using paleoecological techniques to expand the 'time window' of the study, the constraint of the longevity of the individual investigator can be overcome. Further, if paleoecologists can select appropriate systems in which true chronological sequences can be evaluated, they can eliminate the severely restricting and often misleading assumption that observing the differences among communities arrayed across a spatial transect is an adequate substitution for observing long-term changes in time.

1.4 HIERARCHICAL RELATIONSHIPS

Techniques appropriate to approach ecological questions at one scale of resolution may be inappropriate at another; ecological patterns and processes operative at different scales of resolution require different sampling schemes. Ecological systems are fundamentally hierarchical in nature; that is, ecological patterns and processes at a given scale of resolution in space and time are bounded by conditions at higher levels, but include all relationships ongoing at lower levels (Allen and Starr, 1982; Delcourt *et al.*, 1983; O'Neill *et al.*, 1986; Urban *et al.*, 1987). Proper resolution requires a sensitivity to

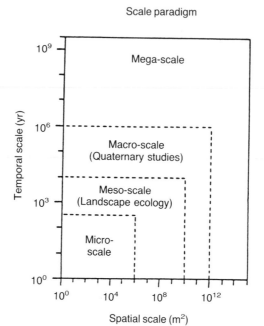

Figure 1.5 Spatial-temporal domains for a hierarchical characterization of environmental forcing functions, biological responses and vegetational patterns. From Delcourt and Delcourt (1988).

these hierarchical relationships and appropriate selection of techniques and development of sample design.

1.4.1 Spatial and temporal scale

Space and time can be scaled usefully within a four-part scheme (Fig. 1.5). As defined here, this scheme includes a **micro-scale domain**, which extends in time from 1 year to 500 years and in space from 1 m^2 to 10^6 m^2; a **meso-scale domain**, from 500 years to 10 000 years and from 10^6 m^2 to 10^{10} m^2; a **macro-scale domain**, from 10 000 years to 1 000 000 years and from 10^{10} m^2 to 10^{12} m^2; and **a mega-scale domain**, from 1 000 000 years to 4.6 billion years and $>10^{12}$ m^2 (Delcourt *et al.*, 1983; Delcourt and Delcourt, 1988). This scaling hierarchy illustrates the expansion in view from short-term, local, plot-specific ecological studies to very broad-scale, long-term evolutionary studies.

The micro-scale domain is most familiar to ecologists who work with contemporary natural or experimental systems; it is the realm of the animal-population and plant-succession ecologist. The meso-scale is the domain of ecologists interested in the development of landscape heterogeneity and the assembly of modern communities and ecosystems, which has occurred over the present interglacial interval. Ecologists interested in the process of species migrations, invasions, and extinctions or major changes in environmental gradients and ecotones across physiographical regions or on a subcontinental scale examine questions relevant to the macro-scale domain of spatial and temporal resolution. Paleoecologists concerned with the effects of major global events on the change in structure and composition of biotic systems over the full span of the Earth's history must use techniques appropriate to resolving changes occurring on the mega-scale domain.

1.4.2 Scaling of organism/population relative to system

In any ecological or paleoecological study the organism or population being studied is most appropriately viewed in a context that includes its life span, life history strategy, recurrence interval and magnitude of disturbances and other environmental factors that affect it, and the spatial and temporal scale over which it operates in the environment and is affected by environmental change (Pickett and White, 1985). To be understood, it must be examined within the appropriate window of space and time, that is, scaled correctly relative to the system in which it functions. The scaling hierarchy (Delcourt *et al.*, 1983; Delcourt and Delcourt, 1988) is a paradigm within which to develop experimental design and with which is emphasized the importance of using techniques appropriate to resolving questions that in turn must be resolvable within the space–time domain at which the questions are directed. A variety of environmental forcing functions acts to make biological change happen, thereby resulting in community patterns that we can observe on the modern landscape or resolve through time (Figs 1.5 and 1.6).

On the micro-scale spatial–temporal domain, discrete disturbances, including wind throw of trees, fire, clear cuts, and activities of bulldozers result in destruction of a vegetation patch, local disruption of nests or burrows, and a physical opening in the landscape mosaic. Accompanying the structural damage is an interruption in the pattern of energy flow and nutrient dynamics, often with accelerated erosion and loss of nutrients. Biotic responses are two-fold. Immediate responses of plant populations include recruitment of new individuals through seeding-in or reinvasion, and establishment of perm-anent seedlings. Longer-term responses include plant succession within the gap. The resulting pattern is that of changes in life-form of dominant plants as the canopy gap closes. In an aggregate of such regenerating disturbance patches, the pattern that results is a shifting-mosaic steady state (Bormann and Likens, 1979) resulting in a relatively species-rich community. Techniques appropriate to studying pattern and process of disturbance patches include direct observation and measurement of immediate changes. Seeding-in can be measured using a combination of techniques including seed bank

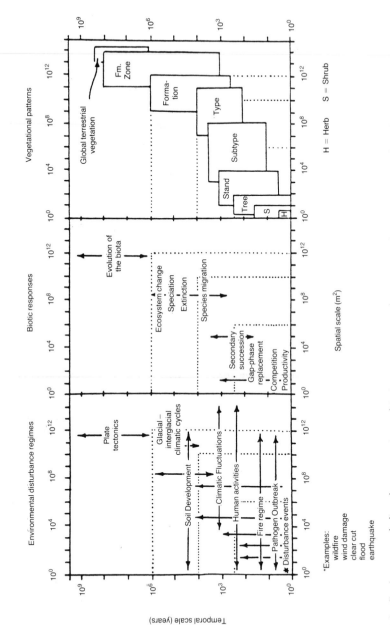

Figure 1.6 Environmental disturbance regimes, biotic responses and vegetational patterns viewed in the context of four space-time domains. From Delcourt and Delcourt (1988).

studies, seed dispersal, and capture–recapture studies of birds and mammals, the potential dispersers of propagules. Inferences concerning the longer-term development of the vegetation mosaic can be accomplished with observations of a number of different patches of known age in different stages of regeneration, in relation to the composition of the general forest mosaic (substituting a spatial array of samples for a temporal sequence in any one plot).

To study the history of forest stand development over a span of 500 years or more, true successional time, requires the use of a combination of historical records – including land survey records, forest inventories, tree-ring analyses, forest-floor excavations, and pollen and plant-macrofossil records from woodland hollows.

The meso-scale spatial–temporal domain is a productive interface between ecology and paleoecology, encompassing events occurring over the 10000 years of the last interglacial interval, the Holocene, and second-order watersheds and larger landscape mosaics (Delcourt and Delcourt, 1988). Important environmental forcing functions include climatic changes and changes in prevalent disturbance regimes and environmental gradients. Biotic responses include changes in community composition across ecotones, invasions and establishment of species as they spread onto newly available landscapes with glacial retreat, and reassembly of Holocene mammal communities following loss of Pleistocene forms. Within the meso-scale domain, human cultural evolution has transformed natural landscapes into cultural ones. Resolvable patterns include changes in vegetation types arrayed along topographic and latitudinal gradients, changes in landscape heterogeneity and patchiness of environments. Techniques appropriate for study are primarily fossil evidence from pollen and plant-macrofossil records from lakes and other wetlands in humid regions, packrat middens such as in the arid western United States, and small mammals analyzed from cave and stream terrace deposits. The archaeological record also provides a wealth of data concerning changes in human populations and settlement patterns.

On the macro-scale, major changes in Quaternary environments and climate have resulted in displacements of biota on a subcontinental scale. Species migrations and changes in the make-up of biomes resulted from both differential rates of spread of plant and animal populations on their leading edges of migration, as well as from local or total extinctions during times of rapid climate change. Landscape changes occur on this scale across entire physiographical regions. Techniques of Quaternary landscape reconstruction include mapping of glacial geology and non-glacial geomorphic changes and mapping of changes in population centers and range margins of plants and animals based upon extensive networks of Quaternary paleoecological sites.

The mega-scale domain includes the entire development of the biosphere, lithosphere, hydrosphere and atmosphere. Major environmental forcing functions operative at this scale include plate-tectonic changes in the configuration of continents and oceans. Patterns of speciation and adaptive radiation as well as extinction characterize the response of the biota to mega-scale geological and climatic events. To detect these patterns throughout the course of the Earth's history, the paleobiologist compiles stratigraphical data concerning the time ranges between the evolution and the extinction of species within major phyla (Nitecki, 1981).

1.5 CONCLUSIONS

1. The field of Quaternary ecology addresses the structure, function and dynamics of populations, communities and ecological systems that existed over the past several million years.

2. The physical, chemical and biological processes existing today have been operative throughout the Earth's history, although they have varied through time in both rates and intensities. We can use our knowledge of modern community composition and the relationships of organisms to their environment as a guide to interpreting fossil plant and animal assemblages as they represent communities that lived in past environments. Therefore, the present is the key to understanding the past.

3. Comparison of modern community composition with communities that existed during the last 20 000 years of the late-Quaternary time interval reveals that although certain plant communities have remained similar in composition through time, some modern communities have no fossil counter-parts, and there have been past communities that lack good modern analogues. This demonstrates that an understanding of the changes in community composition over long periods of time is essential for placing modern communities in proper context. Thus, an understanding of the past is a key to interpreting the present.

4. Techniques appropriate to approach ecological questions at one scale of resolution in space and time may be inappropriate at another scale. Ecological patterns and processes operative at different scales of resolution require different sampling schemes.

5. In any ecological study, the organism or population being analyzed must be viewed within the context of its life span, life history strategy, recurrence interval and type of disturbance regime, and the spatial and temporal scale of the system within which the organism functions.

6. The Quaternary ecological record is particularly useful in providing direct tests of ecological hypotheses for which experimental studies are unavailable because of limitations on human longevity relative to the dynamics of the ecological systems under investigation.

2 Dispersal, invasion, expansion and migrational strategies of populations

2.1 ISSUES

The process of invasion of a species into an area not previously occupied by it is an important facet of population dynamics (Elton, 1958; Harper, 1977; Grime, 1979). In vascular plants, mechanisms of dispersal of propagules, which include spores, seeds and bulbules, range from gravity to wind to animal vectors, including humans (van der Pijl, 1982). Demographic characteristics of plants, including their co-evolved relationships with propagule dispersers, are thus important in determining the efficiency and rapidity with which their populations can spread across a newly available landscape (White, 1985). Population expansion can occur either by vegetative reproduction or by sexual reproduction, resulting in an increase extending radially beyond the point of introduction (Harper, 1977). Dispersal to new environments may occur as a single event, a series of repeated introductions, or several simultaneous introductions from a number of source populations; for successful colonization, dispersal must be followed by establishment of viable populations. The pattern of spread may be a simple diffusion radiating outward from a population center or expanding beyond but parallel to a distributional margin. However, if the landscape represents a heterogeneous mosaic of suitable safe sites and hostile sites for invasion, physical corridors may channel directions of population extension or the establishment of outliers, and barriers may be present to impede the population expansion. Multiple point sources of introduction may result in more rapid expansion of the population than occurs from a single point of introduction (Mack, 1985; Fig. 2.1(a)).

Colonists are plants that are already present in an area and that re-establish within an opening in the vegetation created by disturbance (see Chapter 3). Invasives are species that are newly introduced into an area, either as a result of human activities (Mack, 1985; Mooney and Drake, 1986) or because of a major environmental change such as climatic warming or cooling (Jacobson, 1979; Bennett, 1983; M. Davis et al., 1986). Species may invade either open

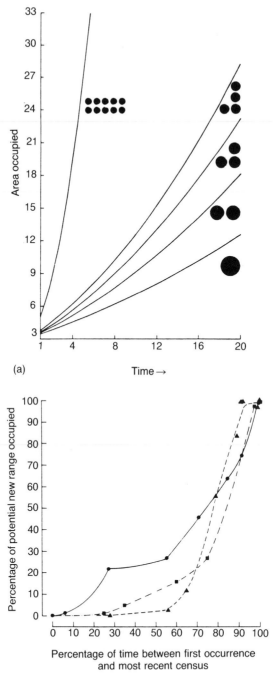

Figure 2.1 (a) The rate of range expansion increases with the number of point sources for introduction of an invading species (indicated by the number of black dots associated with each curve). (b) The historic pattern of invasion for three exotic plant species expressed as a percentage of their potential new range occupied since the time of invasion. From Mack (1985).

landscapes or previously existing plant communities. Once established, the invasive species may affect the community through competitive interactions with the native flora, for example resulting in extinctions (Mooney and Drake, 1986; Crosby, 1986).

Observations on species that have invaded new territory within the last several hundred years because of human introductions have yielded important insights into the patterns and processes of invasion (Elton, 1958; Mack, 1985; Mooney and Drake, 1986). The typical pattern of invasion for vascular plants (Fig. 2.1(b)) involves an initial period of relatively slow spread, which may last for the first 25 to 60% of the time since initial introduction and during which the invasive species may exist only in a few relatively isolated localities. This initial phase in the process of exponential growth is followed by a phase of rapid range expansion. This pattern has been observed for a number of taxa (Fig. 2.1(b)), ranging from perennial grasses (Mack, 1981) to desert shrubs (Robinson, 1965) and cacti (Moran and Annecke, 1979).

Migrations of species result both from their invasions of new habitats beyond one or more margins of their former ranges and from local extinctions of their populations occurring on other parts of their previous distributional ranges (Delcourt and Delcourt, 1987a; Sauer, 1988). If a border of distribution of a plant species is moving, it becomes important to determine whether its population disequilibrium is a result of a change in dispersal capability or in environmental conditions; further, it is important to know whether its leading or advancing edge of migration is rate-limited by the effectiveness of its dispersal mechanisms, by the rate of environmental change, the frequency and area of disturbance in providing suitable sites for occupation, or by biotic

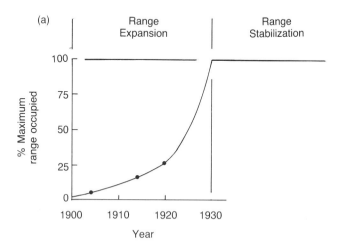

Figure 2.2 (a) Plot of area occupied versus time based on accumulative percentage of eventual range occupied by the exotic grass *Bromus tectorum* invading the American West during the first half of the 20th century. Modified from Mack (1981).

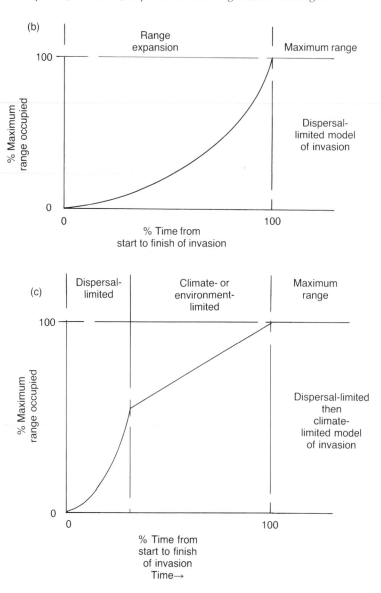

Figure 2.2 (Cont'd)
(b) Hypothetical model for a dispersal-limited pattern of invasion. (c) Hypothetical model for a climate- or environment-limited pattern of invasion.

interactions (M. Davis *et al.*, 1986; Sauer, 1988; Birks, 1989). In many cases, our perception of whether a species is actively migrating is conditioned by the scale at which we observe it (Sauer, 1988).

The case study of the introduction of the non-native species of annual grass, *Bromus tectorum*, into the western United States during the past century illustrates the process of invasion under conditions that permitted widespread population expansion (Fig. 2.2(a)). *Bromus tectorum* was introduced between 1894 and 1899 A.D. into the region of semi-arid steppe vegetation within intermontane basins of Utah, Washington, and British Columbia. Because of human disturbance caused by establishment of settlements along railroads and near mining districts, and because of overgrazing by cattle, sheep, and horses, *Bromus tectorum* was able to invade open sites previously occupied by native grasses. Between 1899 and 1930, this alien grass expanded exponentially in area (Fig. 2.2(a)). As all suitable sites for *Bromus tectorum* became occupied, by 1930 its distributional range stabilized in area as it reached a limit to expansion caused by some environmental, climatic or biotic factor, and the slope of the curve for its rate of expansion changed abruptly (Fig. 2.2(a); Mack, 1981).

The recent history of invasion of *Bromus tectorum* conforms to a dispersal-limited model of invasion (Fig. 2.2(b)). Another possibility is that of climate- or environment-limited invasion (Fig. 2.2(c)), in which a species population initially released from an environmentally or biologically controlled restraint undergoes a period of exponential expansion, followed by a period of slowed rate of expansion that is limited by climate or some other aspect of the environment. Tests of these models can be undertaken using the Quaternary paleoecological record. On a Quaternary time scale, questions of dispersal, invasion and migration have been addressed most effectively for trees using time-series of pollen data from networks of lake sites. Longer-term population dynamics of native trees on watershed, regional landscape and subcontinental scales complement information available from introductions of exotic species occurring over the past several hundred years and allow comparison of natural experiments with human-induced experiments. These data allow us to address the question of whether dispersal is a rate-limiting factor in tree species migrations, or whether climatic change is the 'ultimate ecological control' (Bryson, 1966).

2.2 DISPERSAL

If seeds are dependent solely upon gravity for dispersal, the process of spread of a species may occur very slowly. In the case of recruitment of new individuals of tree species into an opening in an old-field (van der Pijl, 1982), the rate of spread by this mechanism is a direct function of the rate at which trees reach reproductive maturity, which may take 10 to 25 years or longer (Fowells, 1965). Several additional means of effective dispersal of seeds of tree species include (1) water transport, for example in willow (*Salix*) and alder (*Alnus*), which have light seeds that can be carried in swiftly flowing water and colonize stream banks where seeds are deposited; (2) wind transport, for example the seeds of birch (*Betula*), aspen (*Populus*), and conifers such as eastern hemlock (*Tsuga canadensis*), which may be carried airborne for long distances, especially across open, treeless landscapes; and (3) animal dispersal, by mammals, insects or birds.

Alder (*Alnus*) and willow (*Salix*) are documented to have spread across continental Europe at rates of from 1000 to 2000 meters per year during the late-glacial and early Holocene intervals (Huntley and Birks, 1983). Rather than in a pattern of diffusion radiating out in all directions from their Pleistocene refuge areas in southern Europe, the migration routes of these taxa occurred preferentially along the major river systems, which flow predominantly northward and northwestward across central Europe north of the Pyrenees and Alps (Huntley and Birks, 1983). In contrast, in eastern North America the principal river systems flow to the south and southeast, in opposite sense to the general migrational directions for trees recolonizing formerly glaciated landscapes in the Midwestern United States and Canada. Consequently, riverine transport of seeds was probably not a major mode of northward dispersal of plants during postglacial times in eastern North America (Delcourt and Delcourt, 1987a).

The effectiveness of wind dispersal of arctic and alpine plants today has been demonstrated by the study of Glaser (1981), who documented long-distance dispersal of propagules over the surface of ice and snow cover in Alaska. In lower vascular plants, such as ferns, dispersal of spores by wind is probably the most important factor in the cosmopolitan nature of fern floras (Sauer, 1988). An example of long-distance or 'jump' dispersal (Pielou, 1979) that may have been a result of wind transport of seeds is the case of the migration of aspen (*Populus*) into northern New England and adjacent Canada approximately 12 000 years ago (R. Davis and Jacobson, 1985), when an advance colony established as an outlying population located far to the northeast of its main range within a large expanse of treeless tundra. Birks (1989) suggested another possible example of jump dispersal in the case of scots pine (*Pinus sylvestris* var. *scotia*), which exists today as genetically distinctive populations confined to native woodlands in Scotland. Using radiocarbon-dated fossil-pollen sequences from 135 sites across Great Britain, Birks (1989) documented that the initial establishment of *Pinus sylvestris* occurred in northwestern Scotland between 8500 and 8000 years ago. This pine established its northernmost sites in the British Isles by 8000 years ago at a location that was more than 450 kilometers distant from the nearest pine populations in southern Ireland and 350 kilometers from possible seed sources in central Britain. This apparent jump dispersal event for *Pinus sylvestris* may have resulted from the wind dispersal of pine seeds from source populations in southern Britain, from more distant populations in northwestern Europe, or even from localities along the coastal zone of the North Sea basin (Birks, 1989).

In a series of experimental studies in Costa Rica, Janzen (1970; references cited in Janzen, 1983) has demonstrated the importance of large mammals including recently introduced horses (*Equus*) and tapirs (*Tapirus bairdii*) in dispersing heavy-seeded, fleshy fruits that are characteristic of many tropical tree species, and hence to act as important agents in maintaining species richness of tropical forest communities. In the Pleistocene, large mammals, including mastodonts (*Mammut americanum*), horses, and tapirs, were abundant over much of what is today the Temperate Zone of North America (Kurten and Anderson, 1980). Mass extinctions of these 'megafaunal' species occurred at the end of the Pleistocene, by about 10 000 years ago (Martin and Klein, 1984). Some of the now-extinct large mammals may have been important seed dispersers that in previous glacial and interglacial times were effective in maintaining high species richness in both tropical and temperate forests (Janzen and Martin, 1982). It is possible that some of what are today relatively rare temperate trees, such as osage orange (*Maclura pomifera*) and Kentucky coffee tree (*Gymnocladus dioicus*), which have large, fleshy fruits that are not dispersed far beyond the parent tree by gravity alone, were dispersed previously by large mammals that are now extinct. The implication is that some temperate trees that have been rare and localized in distribution during the present interglacial interval may have been much more common and more widespread during previous interglacials.

Dispersal of seeds by small mammals is probably not a major mechanism for the migration of deciduous, mast-bearing trees such as oak (*Quercus*) and beech (*Fagus grandifolia*) during times of major Quaternary climatic change. Squirrels (*Sciurus* spp. and *Tamiasciurus hudsonicus*) seldom carry seeds much beyond the parent tree, and other small mammals that prey on seeds also tend to disperse them only locally (Johnson *et al.*, 1981; Johnson and Adkisson, 1985). Johnson and Adkisson (1985) suggest that the foraging behavior of blue jays (*Cyanocitta cristata*) makes them ideal candidates for rapid dispersal of both oaks and beech. For example, Johnson and Adkisson (1985) observed that in late summer and early fall in southern Wisconsin, jays from a wide area visit oak and beech woods, both eating up to 25% of the nut crop and collecting nuts to cache in the ground upon return to their territories. The average distances travelled on round-trips range from a few meters to over 1 km; the maximum transport of beechnuts by jays over open country was 4 km (Johnson and Adkisson, 1985). When flying, jays can hold 3 to 5 acorns or up to 14 beechnuts at a time. On the average, 16% of acorns cached by jays are deposited within closed forests, with 67% cached in relatively open early-successional forests, and 12% in neighboring grasslands (Johnson and Adkisson, 1985). In western Virginia, Johnson has recorded that in 28 days of observation, 50 jays transported and cached 150 000 acorns from a grove of pin oak trees. Cached seeds are retrieved in the spring, but many survive to sprout and thus disperse the species of oaks and beech widely from the original populations. The net transport by jays represents a flux of acorns away from source oak trees toward suitable sites within open grassland or forest-edge environments (Johnson and Webb, 1989).

This type of selective dispersal must have broad implications for determining the structure of deciduous forest communities, both today and throughout postglacial times. Johnson and Adkisson (1986; Johnson and Webb, 1989) suggest that during the time of postglacial migrations of deciduous trees, blue jays could have nested in young, oakless forest patches, caching nuts collected from mature oaks a few kilometers to the south, and thus could have expanded the northern ranges of oaks progressively into boreal forest or across tundra over the period of postglacial climatic warming. Today, the yearly flight of jays between temperate deciduous and boreal coniferous forests provides a mechanism for the northward dispersal of acorns along postglacial flyways, an example of tree migration facilitated by this plant–bird interaction.

The now-extinct passenger pigeon (*Ectopistes migratorius*) has been proposed as a Holocene seed disperser of deciduous trees. Prior to its extirpation in 1914 A.D. (S. Webb, 1986), migratory populations of the

Table 2.1 Estimates of Holocene migration rates of nut trees

	Beech (Fagus grandifolia)	Oak (Quercus)	Hickory (Carya)
Minimum age of seed production	40 yr	20 yr	30 yr
Average rate of postglacial range extension			
Fossil-pollen data (M. Davis, 1976, 1981a, 1983)	200 m/yr	350 m/yr	200 to 250 m/yr
Quantitative forest composition based on fossil-pollen data (Delcourt and Delcourt, 1987a)	169 m/yr	126 m/yr	119 m/yr
Average dispersal distance			
Based on fossil-pollen data (S. Webb, 1986)	8 km/gen	7 km/gen	6 to 7.5 km/gen
Based on quantitative forest composition as reconstructed from fossil-pollen data (Delcourt and Delcourt, 1987a)	6.8 km/gen	2.5 km/gen	3.6 km/gen

passenger pigeon approached five billion. Sarah Webb examined the evidence that passenger pigeons were consumers and possible inadvertant vectors for dispersal of beechnuts. Initial estimates of Holocene migration rates of nut trees (average distances of spread per year) were based upon migration maps constructed using 62 fossil-pollen sites (Table 2.1) distributed primarily across the Great Lakes region and New England (M. Davis, 1976, 1981a, 1983). Margaret Davis considered that the time of first marked increase in pollen values (percentage or influx) recorded the time of arrival of a tree species' populations near each paleoecological site. Using silvicultural literature, Sarah Webb determined the minimum age for seed production reported for individuals of the tree species. She multiplied the values for average migration rate (expressed in meters per year) by the minimum age of reproduction (representing a tree generation) in order to calculate the average dispersal distance in kilometers per generation required to account for the observed patterns of Holocene range extension. Based upon the migration rates as determined from fossil-pollen data (M. Davis, 1976, 1981a, 1983), the dispersal distances for beech, oak, and hickory were calculated as 8, 7, and 6 to 7.5 kilometers per generation, respectively (S. Webb, 1986; Table 2.1).

Sarah Webb (1986, 1987) reasoned that, as these values of dispersal distance substantially exceeded observed contemporary distances of seed dispersal recorded, for example, for blue jays (Johnson and Adkisson, 1985; Johnson and Webb, 1989), low-probability events of 'jump' dispersal over much greater distances might be necessary to account for prehistoric migration rates. The occasional and inadvertant 'sweepstakes' or jump dispersal of nuts may have resulted from nut consumption and transport by seed predators such as passenger pigeons. This would result in discontinuous range expansion of isolated advance colonies of beech beyond the continuous limit of the northern (leading) edge of its range.

Sarah Webb (1986, 1987) suggested that this mechanism of propagule transport by a bird vector was consistent with a spatially discontinuous pattern of mid-Holocene establishment of invading beech populations across the central and western Great Lakes region. The plant-fossil sequences preserved in lake sites· from Michigan, Wisconsin, Illinois, and Indiana document the long-distance establishment of outlier populations of beech, 'jumping' across physical and biotic barriers. Sarab Webb considered that the colonization by beech populations in isolated sites of eastern and southeastern Wisconsin about 6000 years ago constituted several discrete episodes of dispersal, either across the aquatic barrier of Lake Michigan (requiring a dispersal event of 25 to 120 kilometers from a primary population center in Lower Michigan) or across the intervening filter barrier of prairie and oak savanna, the northeastern margin of the Prairie Peninsula (Benninghoff, 1964; Kapp, 1977), which was a xeric zone of unsuitable habitat blocking beech migration. Jump dispersal of beech across the Prairie Peninsula may have involved local establishment in a series of 'safe-sites' with dispersal events of 25 to 130 kilometers distance (S. Webb, 1986). These possible safe sites might be represented by roost trees for passenger pigeons in riverine corridors of mesic gallery forests, extending through the region of upland prairies.

Fossil bones of both passenger pigeon and blue jay have been recovered with faunas of late-Pleistocene and early-Holocene age, primarily from cave sites in unglaciated North America (Lundelius *et al.*, 1983). For example (Fig. 2.3), late-Quaternary fossil specimens of passenger pigeon have been reported in the region extending from northern Florida to Virginia and west

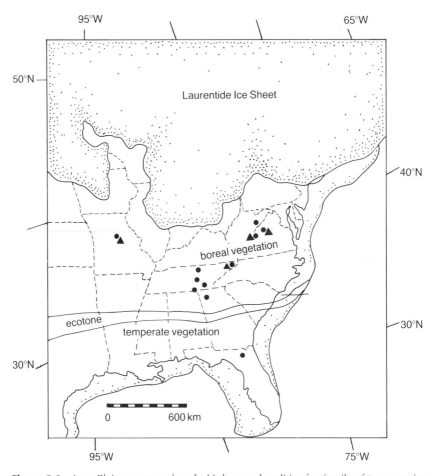

Figure 2.3 Late Pleistocene and early Holocene localities for fossils of two species of birds, passenger pigeon (*Ectopistes migratorius*, dots) and blue jay (*Cyanocitta cristata*, triangles) (data from Lundelius et al., 1983). Late Pleistocene vegetation map for 20 000 to 18 000 yr B.P. From Delcourt and Delcourt (1987a).

to Missouri, across the glacial-age ecotone between southern temperate and more northern boreal vegetation. Late-glacial and early-Holocene localities for fossil blue jays are situated along the trend of the Appalachian Mountains from northeastern Tennessee to northern Virginia and west to the central Missouri Ozarks. The fossil bird remains occur with mammal and plant assemblages that were indicative of the late-glacial changeover from boreal to temperate plant communities, during which populations of a number of deciduous tree species were migrating northward. The paleontological record thus supports the interpretation that birds facilitated the dispersal of nut-bearing tree species during their postglacial migrations.

Subcontinental-scale reconstructions of forest composition have been mapped quantitatively across the eastern half of North America for the last 20 000 years (Delcourt and Delcourt, 1987a). Based upon 162 sites with radiocarbon-dated plant-fossil sequences, late-glacial and postglacial migrations have been reconstructed for a number of important forest trees. Migration rates calculated from these maps along five separate routes (distributed from the continental interior to the Atlantic Seaboard) yield average northward migration rates for beech, oak, and hickory of 169, 126 and 119 meters per year, respectively (Table 2.1). The refinement in calculated values for average migration rates is based upon a broader geographical coverage of paleoecological sites and quantitative forest composition that records the time of first arrival rather than of major population expansion (Delcourt and Delcourt, 1987a). Revised postglacial values for average dispersal distance are 6.8 kilometers per tree generation for beech, 2.5 for oak and 3.6 for hickory. These values are more consistent with those reported by contemporary studies of nut transport by jays (Johnson and Webb, 1989). The average postglacial value of 6.8 kilometers per generation for beech presumably reflects both high-probability transport in short trips (less than 4 km) and occasional lower-probability episodes of jump dispersal (up to 130 km across barriers).

During full-glacial times, 20 000 years ago, small populations of *Fagus grandifolia* trees persisted on favorable edaphic sites distributed across the southeastern United States (Delcourt and Delcourt, 1987a). The subsequent postglacial expansion in range of beech across eastern North America occurred at generally very low population densities (Bennett, 1985; Delcourt and Delcourt, 1987a). Beech became important locally in the southeastern United States during the transition from late-glacial to early-Holocene times, when climatic conditions were cool and moist (Watts, 1980a,b; Delcourt *et al.*, 1983). Only after about 8000 years ago, however, did reconstructed population levels of beech trees reach 20% of the forest composition, primarily in the Great Lakes region (Delcourt and Delcourt, 1987a). This pattern of postglacial spread of beech populations is consistent with the contention of Carter and Prince (1981) that the density of sites necessary for maintaining a population of plants must reach a certain threshold before successful colonization can occur, and that the expansion of distributional limits depends on an increase in the rate of addition of suitable sites relative to the removal of suitable sites over the same time interval. The number of suitable sites that must be available in order for populations to spread is relatively low for long-lived trees because the removal rate of perennial plants is lower than that of annuals (Carter and Prince, 1981). The observed postglacial rate of spread of beech, occurring as a patchwork at relatively low population densities and facilitated by the interaction of seed dispersers such as blue jays, supports this supposition (Bennett, 1985; Johnson and Webb, 1989).

2.3 INVASION

Consideration of dispersal mechanisms is relevant to determining the relative roles of climatic change and dispersal as factors in postglacial forest development in temperate regions. For certain taxa such as spruce (*Picea*), that had their full-glacial refuges close to the southern limit of the continental ice sheets, the limiting factors to northward spread seem to have been the rate of climatic warming coupled with the rate of retreat of the physical barrier of the ice sheet (Bernabo and Webb, 1977; Ritchie and MacDonald, 1986; Delcourt and Delcourt, 1987a). Spruce seeds are light and winged and could be wind-dispersed. However, for warm-temperate trees that in many cases had refuge areas far from the ice sheet and in limited areas, the means and rate of dispersal were at least as important as climate change in determining the rate of advance of their leading edge of distribution. The question is whether dispersal of tree species has been limiting to the extent that it caused a substantial time-delay or lag in their northward invasions into regions otherwise with climatic conditions suitable for their survival (M. Davis *et al.*, 1986).

2.3.1 Dispersal versus climate

In a recent paper by Margaret Davis and colleagues (1986) this question was considered explicitly using the fossil records of American beech (*Fagus grandifolia*) and eastern hemlock (*Tsuga canadensis*), both of which can be readily identified to species either by their fossil pollen or macrofossil remains. Hemlock exemplifies the type of tree that has light, winged seeds that are dispersed by wind, whereas beech has the heavier fruits and nuts that are dispersed by birds and small mammals. Both tree species spread into the Great Lakes region between 7000 and 5000 years ago, and advanced to their northwestern range limits in Wisconsin in the past 2000 years, after the range limits of other temperate trees had largely stabilized throughout this region. Was this continued expansion because of a continued response to climate

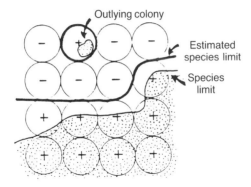

Figure 2.4 Sampling grid of small lakes with paleoecological sequences designed to monitor the timing and geographic pattern of dispersal of a tree species across a landscape. Within the circular source area surrounding each lake, the symbol + means that significant quantities of pollen grains of the tree species are present within the lake sediments, and the symbol − means that pollen of the tree species is absent. From M. Davis *et al.* (1986).

change, or was it a lag effect because of poor dispersal, for example, across the broad expanse of Great Lakes such as Lake Michigan, that previously prevented the species from occupying their potential natural ranges?

To detect the time of establishment of local populations of beech and eastern hemlock based on pollen analysis, a sampling grid was developed that included small lakes (each 3–10 ha) spaced about 50 km apart (Fig. 2.4). Choosing small lake sites restricted the area from which the pollen rain was derived for each site (Jacobson and Bradshaw, 1981) and formed a dense enough network with which to calibrate pollen percentages detected in the near-surface sediments with the known presettlement range margin and outlying colonies for each species determined from land survey records. They found that fossil pollen percentages higher than 1% of hemlock or greater than 0.5% of beech occurring consistently in pollen counts can be considered evidence that the presettlement range limit was located within 20 km of the pollen site (M. Davis *et al.*, 1986).

These relationships were then applied to the fossil records from about 50 radiocarbon-dated lake sites in Michigan and Wisconsin in order to establish the postglacial times of first arrival of beech and hemlock (Figs 2.5, 2.6; percentages plotted at approximate 1000-year intervals). Beech first arrived in southern Lower Michigan about 7000 years ago; by 6000 years ago it had crossed Lake Michigan into southeastern Wisconsin. The population that established in southeastern Wisconsin must have been derived either from dispersal directly across the lake or around its southern end, which at that time consisted of discontinuous patches of woodland growing in fire-protected habitats within the matrix of a prairie landscape (Prairie

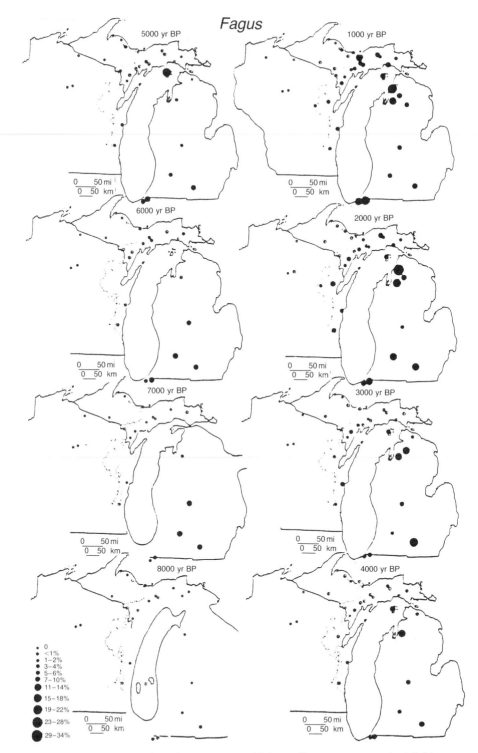

Figure 2.5 American beech (*Fagus grandifolia*) pollen percentages in Michigan and Wisconsin from 8000 to 1000 yr B.P. From M. Davis *et al.* (1986).

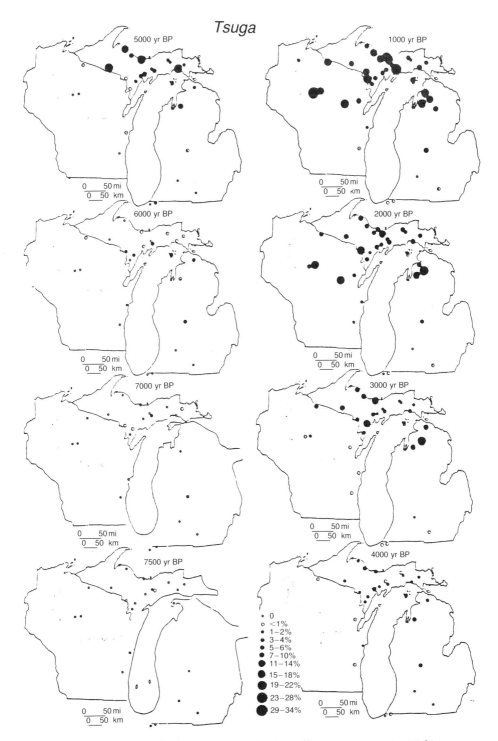

Tsuga

Figure 2.6 Eastern hemlock (*Tsuga canadensis*) pollen percentages in Michigan and Wisconsin from 7500 to 1000 yr B.P. From M. Davis *et al.* (1986).

Peninsula). East of Lake Michigan beech expanded northward, reaching northern Lower Michigan and the Straits of Mackinac by about 5000 years ago, colonizing offshore islands such as Beaver Island in the northern portion of Lake Michigan (Kapp *et al.*, 1969). Low percentages of beech pollen were only discontinuously represented in lake sites in central Upper Michigan between 5000 and 3000 years ago. This appears to reflect a series of dispersal events but with sporadic episodes of establishment and local extinction of beech populations until after their final, successful colonization and population expansion between 3000 and 2500 years ago (Woods and Davis, 1989). Beech explained rapidly west of Green Bay, Wisconsin, only after 2500 years ago (Fig. 2.7). To reach southeastern Wisconsin by 6000 years ago, beechnuts had to be carried over lake barriers 25 to 120 km wide, which would have required a fairly rare long-distance dispersal event by blue jays or passenger pigeons. With the time lag between 7000 and 6000 years ago in the spread of beech from primary populations in Lower Michigan to outlier

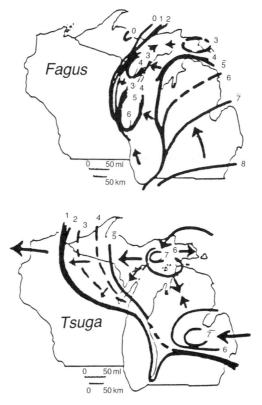

Figure 2.7 Maps depicting the spread of *Fagus* and *Tsuga* across the Great Lakes region during the past 8000 years. Lines show the positions of the species limits with numbers indicating time in thousands of years B.P. From M. Davis *et al.* (1986).

populations in southeastern Wisconsin, the observed 1000-year dispersal lag could have been due to unfavorable climatic conditions or to the presence of Lake Michigan as a migrational barrier. The delay in effective colonization of the Upper Peninsula of Michigan from 5000 to 3000 years ago could have also been due to ineffective seed dispersal (M. Davis *et al.*, 1986; Woods and Davis, 1989), although the distances of water to be crossed were considerably less in northern rather than central portions of Lake Michigan, and several islands such as Beaver Island were available as 'stepping stones' to be colonized across this aquatic barrier. Lakes studied in the eastern portion of the Upper Peninsula of Michigan are situated in areas of sandy glacial outwash and not on sandy-loam till, and may not have recorded presence of local beech populations because of their location on unsuitable sites for beech (Whitney, 1986). Additional paleoecological sites studied from more mesic settings may detect earlier establishment of beech related to soil conditions (M. Davis *et al.*, 1986).

As portrayed in Figures 2.5 and 2.6, postglacial fluctuations in the water level of Lake Michigan were caused by meltwater released by retreating glaciers, by uplift of the land following retreat of the massive continental ice sheet, and by shifting positions of lake outlets, as well as by changes in precipitation (Hansel *et al.*, 1985; Hansel and Mickelson, 1988). These dynamic Holocene changes in lake area may have influenced the effectiveness of seed dispersal across this aquatic barrier. Between 10 000 and 5000 years ago the level of Lake Michigan dropped by as much as 61 m (below modern levels) during the low-water Chippewa lake phase (Hansel *et al.*, 1985). This exposed new islands in the central basin of Lake Michigan and may have permitted beech to island-hop from Michigan's mainland to the Wisconsin mainland in a series of jumps, each 30 to 50 km, between 8000 and 7500 years ago (Fig. 2.5). However, by 6000 years ago, the establishment time for first beech arrival in Wisconsin (S. Webb, 1986), the water barrier was 80 to 120 km broad. High-water stages of Nipissing and early Algoma lake phases (between 5000 and 3300 years ago) represented a maximum rise of water level up to 7 m above the modern level of Lake Michigan. High lake levels between 5000 and 3300 years ago would have inundated many of the smaller islands in the Straits of Mackinac and flooded low-lying coastal areas of both peninsulas of Michigan. The corresponding distances for beech dispersal remained on the order of 10 km to 40 km between islands or mainlands across the Mackinac Strait between Lower and Upper Michigan. The distance over which beech dispersed successfully between Wisconsin and Lower Michigan 7000 to 6000 years ago (80 to 120 km) far exceeded the more limited distances (10 to 40 km) between Upper and Lower Michigan 5000 to 3000 years ago that proved an effective barrier to later beech migration. Thus, climatic and environmental conditions in Upper Michigan, rather than dispersal capability, must have limited the rate of spread of beech in the late-Holocene interval (M. Davis *et al.*, 1986; Woods and Davis, 1989).

Between 7000 and 6000 years ago pollen grains of eastern hemlock appeared in several sites scattered across both Lower and Upper Michigan (Fig. 2.6), revealing that the species established in a series of small, isolated outlier populations, perhaps reflecting long-distance dispersal of winged seeds from a primary population center at least 150 km away, farther to the east and southeast in southern Ontario north of Lake Huron (M. Davis *et al.*, 1986). Once established within the central Great Lakes region, eastern hemlock populations expanded rapidly, colonizing the eastern half of Upper Michigan between 6000 and 5500 years ago. During the next 5000 years (Fig. 2.6), westward expansion occurred more slowly across western Upper Michigan and northeastern Wisconsin. Based upon this paleoecological evidence for eastern hemlock, dispersal appears to have been the limiting factor to its initial establishment and expansion across the Great Lakes region rather than climate. Subsequently, during the same time that eastern white pine (*Pinus strobus*) was expanding westward across Minnesota and that the Prairie Peninsula was receding to the west (Jacobson, 1979), slower westward expansion of hemlock may have been controlled by regional climatic boundaries (Bartlein *et al.*, 1984).

During the Holocene interglacial, both climatic change and rates of dispersal must be considered as factors influencing the patterns of establishment and spread of tree species such as beech and eastern hemlock. In the late-glacial and early-Holocene intervals, the rate of climatic and paleoenvironmental change was fast enough that in some cases species responses lagged sufficiently to indicate that dispersal may have been limiting to their range expansions in regions of potentially suitable climate; subsequently, migration lags during the mid- and late-Holocene intervals that resulted from dispersal limitations have been insufficient in most cases to distinguish this effect from response to gradual climatic change (M. Davis *et al.*, 1986).

By 5000 years ago, beech and eastern hemlock had dispersed into and established tree populations within the central Great Lakes region. Both tree species are shade-tolerant and fire-sensitive and become canopy dominants of late-successional forests on mesic sites that are relatively free of disturbance. Whitney (1986) examined original survey records of the General Land Office dating from circa 1836 to 1859 A.D. to characterize the presettlement distribution, composition, and dynamics of hemlock–beech and beech–sugar maple–hemlock forest communities in 32 townships of Crawford and Roscommon Counties, located in the northern portion of the Lower Peninsula of Michigan. Eastern hemlock–beech forests occupied flat to rolling uplands, with underlying substrates of fine-grained loams of glacial tills and clayey lacustrine deposits and medium-grained kame (glacial) deposits (Fig. 2.8). Beech and sugar maple (*Acer saccharum*) occupied both upland and lowland habitats with clayey to sandy loam soils where abundant soil fertility and moisture favored their growth. The frequency of

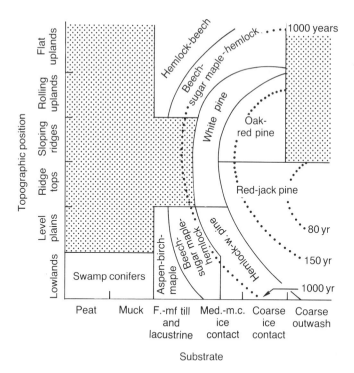

Figure 2.8 Vegetation diagram for presettlement forests of Michigan indicating the distribution of forest types relative to substrate, topography and recurrence interval of disturbance. From Whitney (1986).

observations by American land surveyors of windthrow and burned areas along the survey transects provided estimates for recurrence intervals of natural disturbance in presettlement hemlock–beech forests. In mixed, mid- to late-successional forests dominated by eastern hemlock, beech, and other northern hardwoods, typical return time for catastrophic windstorms ('return time' representing the span of time required for a type of disturbance to impact an area equivalent to that of the forest type in the study area) occurred at intervals of at least 1200 years, with an average recurrence interval of at least 1400 years for severe crown fires. As wind damage and burned areas were reported by surveyors from mutually exclusive areas, the combined extent of physical damage through blowdown and fire probably would be better approximated by an average recurrence interval of about 650 years (Fig. 2.8; Whitney, 1986). Prior to the arrival of EuroAmerican settlers, in the 5000 years since invasion of beech and hemlock into the region, no more than seven or eight cycles of recurrence intervals of disturbance have occurred in mesic forests dominated by eastern hemlock and beech in northern Lower Michigan.

2.3.2 Long-term patterns of invasion

The patterns of invasion recognizable over the past hundred years for introduced invasives (Fig. 2.1) may differ appreciably from those that occurred during postglacial migrations following global climatic warming. Intercontinental transport of weed seeds by humans, whether intentional or inadvertent, represents a situation very different from postglacial invasions, both in the duration of the phenomenon and because of the context into which the species are introduced. The invasive species introduced from a different continent may be ill-adapted to compete with native plants, or on the other hand, the native species may be outcompeted by it (Mooney and Drake, 1986). The fossil record offers a test of the general applicability of theory of invasions based upon patterns and processes observed for recent invasive species (Mack, 1985; Birks, 1989).

Data relevant to examining rates of expansion in area since the last full-glacial interval, 20 000 years ago, are available for important forest trees of eastern North America (Delcourt and Delcourt, 1987a). These maps of reconstructed forest composition are based upon fossil-pollen data, calibrated using relationships of pollen productivity and dispersal determined from an extensive geographical array of modern pollen samples and forest inventory data (Delcourt *et al.*, 1984). The paleo-population maps depict changes in range margins and population centers of trees on a backdrop of changing geographical setting, including locations of proglacial and postglacial lakes, changes in shorelines of the Atlantic Ocean and Gulf of Mexico, and changes in position through time of the extent of the Laurentide Continental Ice Sheet (Delcourt and Delcourt, 1987a).

The dimensionless plot devised by Mack (1985) compares species with different times of introduction and different life-history characteristics by plotting the percentage of potential new range occupied against the percentage of time elapsed between first occurrence and most recent population census (Figs 2.1 and 2.2). This type of graph can also be used to examine the relative rates of expansion of trees during postglacial times (Figs 2.2 and 2.9). Postglacial migrations of boreal and temperate trees occurred over more than 15 000 years. Rescaling to a plot of percentage time elapsed since the onset of invasion versus percentage of ultimate new distributional range occupied eliminates differences that might appear as a function of scaling differences in the organism/system relationship. This is thus a straightforward way to compare the fossil and modern systems directly to see if general principles emerge. In this comparison it is important to plot data for tree taxa that are recognizable to species level in the fossil-pollen record, in order to eliminate complications that would arise from using genera with many species of widely differing ecological amplitude. In the following examples of individual plant species the only exception to this detailed level of taxonomic resolution is made in the case of the plant-fossil evidence for

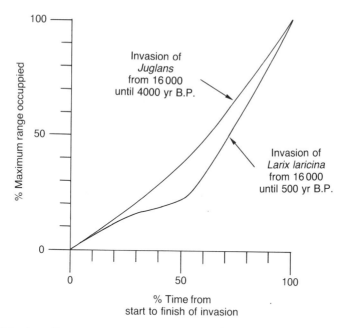

Figure 2.9 Late-Quaternary range expansions of tamarack (*Larix laricina*) and walnut/butternut (*Juglans nigra* and *J. cinerea*) in eastern North America. Data from Delcourt and Delcourt (1987a).

walnut/butternut (*Juglans nigra/Juglans cinerea*), which includes two ecologically similar species (Fowells, 1965).

Suitable full-glacial refuge sites for tamarack (*Larix laricina*), a boreal species today characteristic of acidic, bog environments across central and eastern Canada, were sparsely distributed across unglaciated eastern North America (Delcourt and Delcourt, 1987a). The plot of expansion in area for tamarack (Fig. 2.9) illustrates a long time for initial expansion. After 26% of time since initial expansion had elapsed (by 12 000 years ago), tamarack occupied only 15% of its modern range, and by 52% of elapsed time (8000 years ago), it was still only covering 23% of what is today its natural distributional area. However, rapid expansion occurred between 8000 years ago and 500 years ago (from 52% to 100% of time elapsed since the start of its northward advance). The shape of this curve reflects exponential growth of populations and strongly resembles the curves presented by Mack of recent invasives (1985; Figs 2.1 and 2.2(a)). During the late-glacial interval, tamarack probably relied on chance dispersal to suitable sites, which were widely scattered, resulting in a long time for initial expansion in area. After 8000 years ago, contiguous bog habitat became increasingly available in the northern portion of its range as poorly drained depressions left by the receding glacier filled in and peatlands coalesced as areally predominant features of the landscape. This greater contiguity of habitat would have facilitated a rapid expansion in its range relatively late in the migration history of tamarack (Delcourt and Delcourt, 1987a).

The cumulative expansion curve for walnut/butternut (Fig. 2.9) is also similar in shape to that of recent invasives documented by Mack (1985; Figs 2.1 and 2.2(a)). After 33% of their migration time had elapsed (by 12 000 years ago), walnut and butternut together still occupied only 23% of their eventual postglacial area (Fig. 2.9); however, this was followed by rapid expansion, at 67% of elapsed time (by 8000 years ago) attaining 56% of their late-Holocene area, and with expansion continuing thereafter. *Juglans* was probably confined during full-glacial times to small and widely scattered pocket refuges in the southeastern United States (Delcourt and Delcourt, 1987a). The slow initial expansion of walnut and butternut populations may reflect poor dispersal, dependent primarily upon gravity and small mammals.

The full-glacial distribution of balsam fir (*Abies balsamea*) occupied a latitudinal belt primarily confined to the southern portion of what was then the boreal forest, reaching up to about 10% of the forest composition between 34°N and 36°N across eastern North America 20 000 years ago (Delcourt and Delcourt, 1987a). Fir began to move northward with the first climatic amelioration, both northward to the southern margin of the Laurentide Ice Sheet west of the Appalachian Mountains, and at lower elevations along the Appalachian Mountain chain. Along with aspen (*Populus*) and spruce (*Picea*), balsam fir expanded northward following the retreat of the ice sheet, and then it expanded throughout the Maritime Provinces of eastern Canada (Delcourt and Delcourt, 1987a). By the time 21% of its migration time had elapsed, fir occupied 20% of its potential new range; after 62% of its migration it expanded into 62% of its eventual range. By 82% of the time elapsed since fir migration began (about 4000 years ago), *Abies balsamea* had occupied 98% of its total late-Holocene distributional range (Fig. 2.10). The invasion curve for balsam fir fits the behavior expected on the basis of observations of recent invaders (Fig. 2.2 (a) and (b)), with biological controls over its range expansion continuing for the majority of postglacial time, until *Abies* populations saturated all suitable sites. Only in the last 4000 years (the last 18% of the time since fir began its range expansion) has climate been responsible for stabilizing the maximum, modern distributional limits of *Abies*. Rapid and sustained northward spread of balsam fir throughout postglacial time may reflect the absence of appreciable barriers (the Appalachian Mountains served as a corridor of suitable habitat) as well as its position as an advance invader of open ground previously occupied only by tundra or polar desert.

The invasion curve for beech (*Fagus grandifolia*) does not fit the expected curve for strictly biological controls over range expansion (Fig. 2.2 (b)). The change in the slope of the curve for expansion in area of beech through the Holocene (Fig. 2.10) instead may reflect a shift from biological to

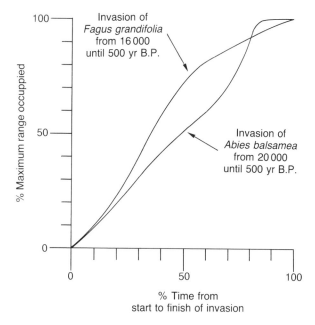

Figure 2.10 Late-Quaternary range expansions of balsam fir (*Abies balsamea*) and American beech (*Fagus grandifolia*) in eastern North America. Data from Delcourt and Delcourt (1987a).

environmental or climatic controls that have predominated for the last 48% of the time since the beginning of beech expansion (Fig. 2.2 (c)). Like walnut/butternut and tamarack, this late-successional, mesic species survived the full-glacial interval in pocket refuges across the southeastern United States (Bennett, 1985; Delcourt and Delcourt, 1987a). Unlike walnut/butternut and tamarack, however, the first expansion of beech was rapid, reaching 32% of its new range by 26% time into migration, and 75% of new area after an elapsed migration time of 52% (by about 8000 years ago). Expansion of beech continued throughout postglacial times (Delcourt and Delcourt, 1987a; Fig. 2.10), slowing dramatically in the middle-Holocene interval as it neared its climatically controlled distributional limits (Bennett, 1985; M. Davis *et al.*, 1986). The initial invasion pattern may reflect the effectiveness of dispersal of beechnuts by jays, nesting in advance of the main population of beech in oak woodlands or mixed northern hardwoods and boreal forest communities (Johnson and Adkisson, 1986; Johnson and Webb, 1989). Beech populations were apparently able to invade directly into relatively species-rich deciduous and mixed deciduous–coniferous forest (Johnson and Webb, 1989).

Eastern hemlock (*Tsuga canadensis*), a wind-dispersed, cool-temperate conifer, experienced a rapid initial expansion (Fig. 2.11) from a very small

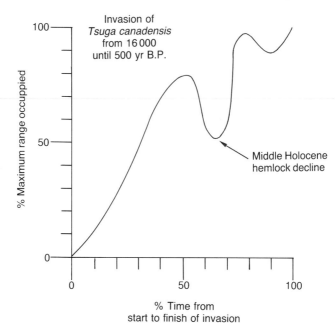

Figure 2.11 Late-Quaternary range expansion of eastern hemlock (*Tsuga canadensis*) in eastern North America. Data from Delcourt and Delcourt (1987a).

refuge area in the southeastern United States (Delcourt and Delcourt, 1987a). By 52% elapsed time since initial northward invasion (about 8 000 years ago), eastern hemlock occupied 80% of its potential new area (Fig. 2.11). However, unlike any other tree species examined (Figs 2.9 and 2.10), the invasion curve for eastern hemlock illustrates a major anomaly in its distributional history, a decline in both dominance and size of range that occurred about 4800 years ago (Fig. 2.11). In the other species examined, rate of expansion of range either accelerated toward the end of their interval of migration or else decelerated slightly. In the case of *Tsuga canadensis*, biological interference in the form of a pathogen has been invoked to explain the mid-Holocene hemlock decline (M. Davis, 1981b).

2.3.3 Genetic implications of the long-term process of invasion

To what extent has the long-term pattern of dispersal and invasion shaped the geographical patterns of variability that we see today in the genetic and morphological characteristics of contemporary tree populations? Can some of the differences in genetic structure (for example, allozyme variation) from central to marginal populations across a species distributional range be appropriately interpreted as the consequence of their postglacial migrational history? To what extent can geographical patterns of genetic variation be explained by a history of invasion with multiple and successive events of long-distance dispersal and establishment of outlier populations?

Cwynar and MacDonald (1987) utilized evidence from both population genetics and paleoecology to evaluate these questions. They reasoned that, with postglacial release of temperate and boreal species from southerly glacial-age refuges, wind-pollinated and wind-dispersed, early-successional plants would expand their distributional ranges northward, as the iterative consequence of long-distance dispersal of propagules and establishment of new, restricted populations, that would in turn provide the seed source for the next dispersal and invasion event. Theoretical studies of population genetics proposed that such events would not necessarily reduce the overall level of genetic variation (reflected by mean heterozygosity), although limited population sizes associated with these founding populations would lead to substantial long-term reduction in allelic diversity (expressed as mean number of alleles per locus; Nei *et al.*, 1975). Cwynar and MacDonald (1987) developed a paleoecological research design to test two hypotheses. For the first hypothesis concerning genetic variability, the reduction in allelic diversity would presumably be most apparent in populations along the periphery of the migrating distributional range, populations established by the most recent event of invasion. They proposed to test the hypothesis that genetic variation in the species populations would systematically decrease in the postglacial direction of migration, directly related to the length of time available for continued population invasion since the first establishment at a site. The length of time since initial arrival at a site provides a proxy for the increased genetic complexity associated with the successive arrival of new populations and enhanced gene flow. To examine the influence of population history upon geographic heterogeneity expressed in plant morphology, Cwynar and MacDonald (1987) reasoned that wind transport would preferentially favor long-distance dispersal of morphologically suitable propagules such as the winged seeds of lodgepole pine (*Pinus contorta*). They hypothesized that with repeated long-distance dispersal events and invasion, directional natural selection in seed morphology would enhance the ability of winged seeds to be transported by wind. For populations along the shifting distribution perimeter, the ongoing process of migration would select for individual trees that reproduced with lighter seeds with a greater wing area relative to overall seed size.

MacDonald and Cwynar (1985) and MacDonald (1987) examined the plant-fossil evidence for the postglacial migration of lodgepole pine (*Pinus contorta* subspecies *latifolia*) from 20 small lake sites distributed from 50° to 65°N across the Western Interior of Canada. This encompassed the 2200-kilometer length of the postglacial migrational route for lodgepole pine, which, over the last 12 000 years, spread northwestward from its ice-age refuge and arrived at its present northern limit in the central Yukon Territory

(a)

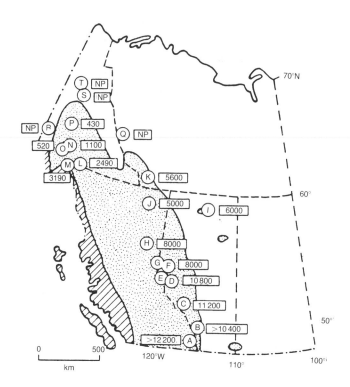

(b)

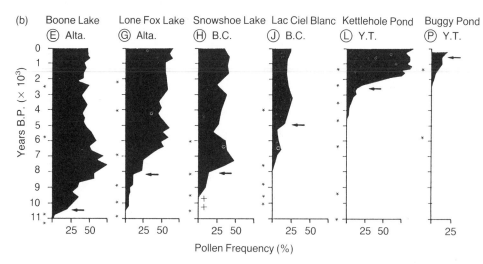

Figure 2.12 (a) The modern range of lodgepole pine, *Pinus contorta* ssp. *latifolia* (dotted pattern) and ssp. *contorta* (diagonal line pattern), in western Canada, and arrival dates in yr B.P. of *Pinus* in the fossil records of lake sites. NP means *Pinus* pollen was not present in the record. (b) Percentages of *Pinus* pollen in six representative paleoecological sites. Arrows indicate the date in yr B.P. at which pine pollen reached 15%, marking its arrival at the site. Asterisks show positions of radiocarbon dates. From MacDonald and Cwynar (1985); reprinted with permission of the National Research Council of Canada.

within the last century (Fig. 2.12(a)). Based upon regional studies of contemporary pollen rain and lodgepole pine distribution, the threshold of 15% pollen of lodgepole pine conservatively records the presence of its population on the landscape (MacDonald and Ritchie, 1986). When the pollen values of lodgepole pine exceed 15% for the first time in a fossil-pollen sequence, it reflects the timing for the first establishment of lodgepole pine on the watershed around a lake site. Figure 2.12(b) illustrates the timing for the initial northward invasion of lodgepole pine and its subsequent population expansion as it successively spread northward through Alberta and the Yukon Territory during the Holocene (MacDonald and Cwynar, 1985). The long-term invasion process for lodgepole pine involved the long-distance dispersal and establishment of small outlier populations, extending as much as 70 kilometers beyond the northern migration front of continuous lodgepole pine populations (Cwynar and MacDonald, 1987).

Today, the most abundant subspecies (*latifolia*) of lodgepole pine is widespread in subalpine forests of western North America, forming even-aged stands on well-drained sites that have been recently burned (MacDonald, 1987). Recent biogeographic studies of lodgepole pine document differences in allelic diversity (Wheeler and Guries, 1982a) and morphological differences in its reproductive structures (seeds, seed wings, and cones; Wheeler and Guries, 1982b). These characteristics of contemporary lodgepole pine populations are correlated with their time since initial invasion of a watershed, based upon their arrival time registered in the fossil-pollen sequence of nearby lake sites.

The paleoecological and genetic evidence supports both sets of hypotheses concerning the population history and ecology of lodgepole pine (Cwynar and MacDonald, 1987). The overall genetic variability (mean heterozygosity, $r^2 = 0.032$, $P = 0.528$) is not correlated statistically with time since arrival (Fig. 2.13(a)), indicating no significant difference observed among populations established at different times (a finding consistent with the theoretical conclusion of Nei *et al.*, 1975). The plot of allelic diversity in Figure 2.13(b) portrays the significant ($P = 0.005$), positive correlation ($r^2 = 0.468$) for the number of alleles per locus in populations with their time elapsed since initial invasion of a watershed. The most recent arrivals near the present northern distribution boundary of lodgepole pine correspond with relatively low allelic diversity, reflecting their long postglacial migration and the cumulative result of genetic bottlenecks associated with repeated long-distance dispersal and successive establishment of small outlier populations. Figure 2.13(c) illustrates the significant ($P < 0.001$), positive correlation ($r^2 = 0.580$) of seed mass and time since arrival; directional selection of propagules has favored lighter seed mass for populations established within the last few thousand years by long-distance events of dispersal. Correspondingly, the ability of a winged seed to disperse can be quantified by the index of the square root of wing loading (Guries and Nordheim, 1984), with the mass of

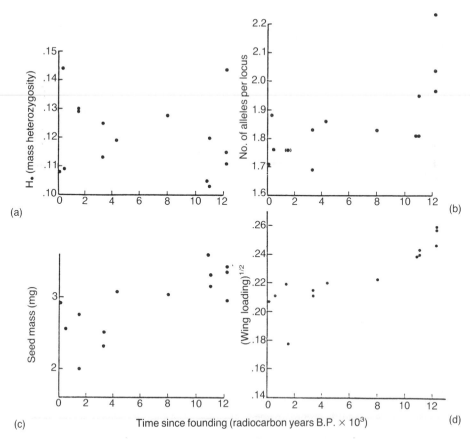

Figure 2.13   Genetic and morphological attributes of populations of *Pinus contorta* ssp. *latifolia* in relation to the time since founder populations were established. From Cwynar and MacDonald (1987).

the seed divided by the surface area of the seed-wing membrane. Lodgepole pine seeds with low wing loading, that is, with large wings relative to their mass, are more readily dispersed by wind currents potentially over longer distances. Figure 2.13(d) supports the conclusion that the migration process of repeated long-distance dispersal, associated with multiple, successive events establishing new outlier populations, favored the selection of seed morphologies with enhanced dispersability by wind (Cwynar and MacDonald, 1987).

Thus, the dispersal pattern of the long-term invasion process and the postglacial sequence of initial establishment and subsequent population expansion can directly account for the relative importance of natural selection in shaping geographical patterns of genetic and morphological variation observed in naturally occurring forest populations.

2.4 EXPANSION OF POPULATIONS

Observing the process of population expansion in long-lived species such as forest trees is virtually intractable using standard experimental approaches of population ecology (Bennett, 1983; Begon and Mortimer, 1982). Fossil-pollen data can potentially give valuable insights into tree population dynamics over long time scales, but their use depends upon calibration of either relative abundances or absolute accumulation rates to changes in tree population sizes on the landscape (Watts, 1973). Most palynological data are reported as changes in percentages of constituent types of pollen grains and spores (Faegri and Iversen, 1975). The percentage constraint prevents their use in this case as a measure of absolute density or dominance of plant populations. In lake environments, pollen accumulation rates (PAR; also referred to as pollen influx; Fig. 2.14) theoretically measure the contribution of each plant taxon independently of the others and may be considered to reflect changes in productivity of the taxon within nearby vegetation (M. Davis, 1963; 1969a, b). Changes in pollen productivity may be caused by changes in overall vegetation density and therefore in the available sunlight influencing flowering and seed set; they may also be related to the dominance of the plant taxon within the plant community (its population size; M. Davis *et al.*, 1973). Although pollen accumulation rates may be affected by the sedimentary environment, which may distort the original pollen influx to the lake, under certain circumstances enough factors can be controlled to examine pollen accumulation rates independent of changes in rates of sediment deposition. The ideal study site for pollen accumulation rates is a closed lake basin that has uniformly continuous sediment accumulation rates

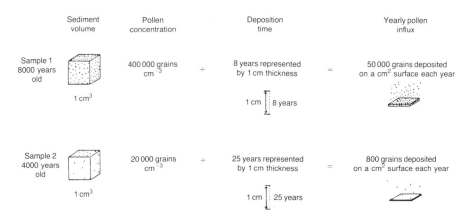

Sediment volume	Pollen concentration		Deposition time		Yearly pollen influx
Sample 1 8000 years old 1 cm³	400 000 grains cm⁻³	÷	8 years represented by 1 cm thickness 1 cm | 8 years	=	50 000 grains deposited on a cm² surface each year
Sample 2 4000 years old 1 cm³	20 000 grains cm⁻³	÷	25 years represented by 1 cm thickness 1 cm | 25 years	=	800 grains deposited on a cm² surface each year

Figure 2.14 Use of pollen concentration data and sedimentation rates to calculate pollen accumulation rates. Modified from Davis (1969b). Reprinted by permission of *American Scientist,* journal of Sigma Xi, The Scientific Research Society.

and that lacks stream inlets and outlets that may introduce pollen grains washing in from soil surfaces or that may remove pollen grains from the lake system (Birks and Birks, 1980).

Case studies of the use of pollen accumulation rates in evaluating rates of increase in tree populations during postglacial time are available from recent studies in England (Bennett, 1983; Bennett and Lamb, 1988); Australia (Walker and Chen, 1987); east-central North America (Delcourt and Delcourt, 1987a); northwestern North America (Tsukada, 1982a); and Japan (Tsukada, 1982b). These studies illustrate the potential for future refinements of techniques in Quaternary palynology to make substantial contributions to understanding the process of population increase in invading populations.

The sequence of invasion of temperate trees onto formerly glaciated landscapes of the United Kingdom during postglacial climatic warming and their subsequent population increases are exemplified by the study of Bennett (1983). At Hockham Mere, Norfolk (a lake basin 1 km long × 0.75 km wide) Bennett (1983) examined a detailed time series of pollen samples from sediments spanning the early-Holocene interval, during which the major tree taxa immigrated from the mainland of central and northwestern Europe to the British Isles. For each tree taxon, which was generally represented by only one or two species, he calculated the time of first increase in PAR after arrival and establishment of populations on the local watershed. PAR values were plotted for the interval over which each tree species showed a sustained increase in pollen accumulation rates. From these plots (Fig. 2.15) Bennett was able to calculate parameters for exponential and logistic growth curves for each tree type and to estimate population doubling times (Bennett, 1983).

The earliest age at which most of the temperate British tree species reproduce ranges from 10 to 50 years after seed germination (for *Pinus sylvestris*, *Quercus petraea*, *Betula pendula*, *Fraxinus excelsior*, *Ulmus glabra* and *Alnus glutinosa*); some oak species do not produce fruit until age 70 to 80 in a closed canopy (Bennett, 1983). The longevity of most of the tree species is as low as 100 years; *Pinus sylvestris*, *Quercus* and *Ulmus*, however, may reach 200 to 300 years of age, and *Tilia cordata* may re-sprout from its base and may persist in place as a multi-stemmed individual for many hundreds of years (Bennett, 1983). Bennett found that, in the fossil record from Hockham Mere, the population expansions occurred over a span of 280 of 1140 years, an interval encompassing many generations of trees. Population doubling times varied from 35 to 175 years (Figs 2.15 and 2.16(a)), from which Bennett suggested that expansions of tree populations were not limited by environmental conditions. *Betula* was the first to increase, followed by *Pinus* and *Corylus*, then the first expansion of *Quercus*, then *Ulmus*, all increasing rapidly during the interval between 9750 and 9250 years ago. A second expansion of *Quercus* (probably a second species) began after 8700 years ago, with a longer doubling time than previous invaders, either reflecting a different set of life-history characteristics or a slower rate of maturity because

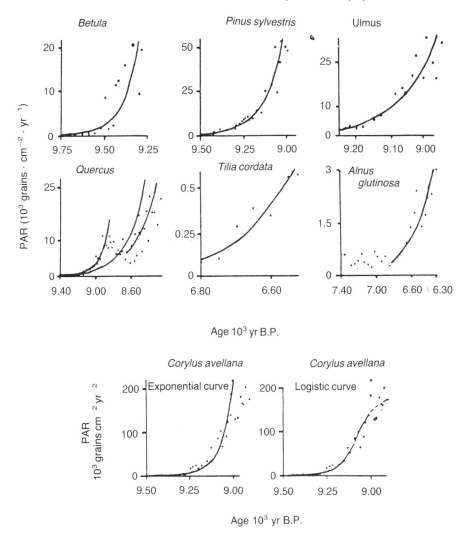

Figure 2.15 Pollen accumulation rates (PAR) during the prehistoric time interval of invasion of representative tree taxa at Hockham Mere, southeastern England. Exponential curves are fitted to the PAR values, and both exponential and logistic curves are presented for *Corylus avellana* (Bennett, 1983). From Delcourt and Delcourt (1987a).

of dense shading by the closed canopy of previously established tree populations (Bennett, 1983). *Alnus* and *Tilia* arrived and expanded their populations at Hockham Mere after 6800 years ago, much later than the other species (Fig. 2.16(a)). These species increased relatively slowly on the watershed of Hockham Mere, with average population doubling times of 99 years for *Tilia* and 174 years for *Alnus*.

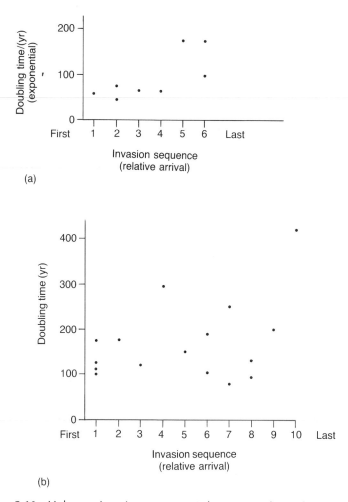

Figure 2.16   Holocene invasion sequence (relative time of arrival) for (a) temperate trees in southeastern England, data from Bennett (1983), and for (b) tropical trees in northeastern Australia, data from Walker and Chen (1987).

Results of studies of the rate of postglacial population increase in temperate tree species in east-central North America (Delcourt and Delcourt, 1987a), the Pacific Northwest of North America (Tsukada, 1982a), and central Japan (Tsukada, 1982b) are consistent with those of Bennett (1983) in concluding that late-arriving, longer-lived, late-successional tree species appear to have slower rates of increase and longer doubling times than those of the earliest arriving, pioneer taxa. One interpretation is that the pre-existing vegetation at some point begins to offer resistance to further immigration and rapid expansion of new arrivals (Smith, 1965), either through saturation of habitats on available environmental gradients or through formation of a closed canopy under which the later arrivals are suppressed, resulting in longer time to flowering and fruiting (as may have been the case for the second species of *Quercus* to arrive in southern England). A second possible interpretation is that later arrivers (e.g., *Quercus*, *Fagus*) may be inherently slower to mature and expand their populations than first arrivers (e.g., *Populus*, *Picea*).

Postglacial synthesis of tropical rainforest has been studied quantitatively by Walker and Chen (1987) on the basis of palynological studies of sediments from Lake Barrine, a volcanic crater lake on the Atherton Tableland of northeast Queensland, Australia (Fig. 2.17). Lake Barrine today is surrounded by tropical rainforest but is located close to the 1300 mm limit of annual precipitation that marks the western boundary of rainforest in Australia (Webb and Tracey, 1981). Lake Barrine has a surface area of 1 km², with no major inflowing streams and a relatively constant depth of 67 m which is maintained by an outlet that is effective during wet summers (Walker and Chen, 1987). Organic-rich clay sediments of Lake Barrine are 6 m deep and date from 10 000 years ago to the present; the fossil charcoal

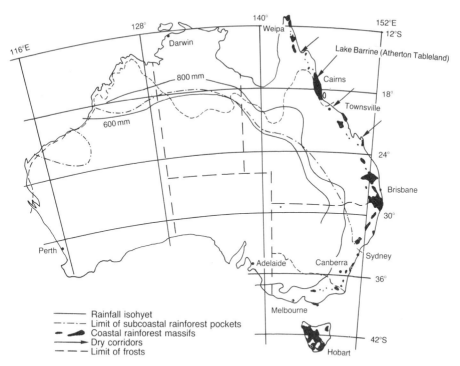

Figure 2.17   Location of Lake Barrine on the Atherton Tableland of northeastern Australia in relation to the modern distribution of Australian coastal rainforest vegetation (black patches). Modified from Webb and Tracey (1981).

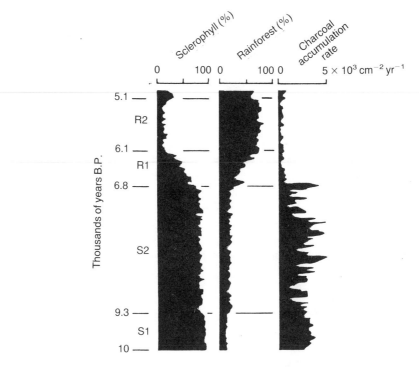

Figure 2.18 Changes in relative frequencies of pollen of plant species characteristic of sclerophyll woodland and rainforest during the early to middle Holocene at Lake Barrine in northeast Queensland, Australia. The decrease in accumulation rates of charcoal in the sediments of Lake Barrine after 6800 yr B.P. reflects diminished fire frequency. Modified from Walker and Chen (1987); reprinted with permission from Pergamon Press.

record (Fig. 2.18) indicates that high fire frequency and intensity characterized the region through the early-Holocene interval. Fire-adapted sclerophyllous vegetation was replaced by tropical rainforest beginning 6800 years ago, following an abrupt decline in fire frequency caused by a climatic change to increased precipitation (Fig. 2.18). This mid-Holocene timing for invasion by tropical rainforest is consistent with other paleoecological studies of the Quaternary history of crater lakes on the Atherton Tableland (Kershaw, 1981).

Pleistocene refuge areas for tropical rainforest species were located to the east of Lake Barrine and were small and isolated along the Australian coast; modern patterns of rainforest distribution still show disjunctions in tropical species ranges (Fig. 2.17). The sequence of postglacial immigrations of tropical rainforest species shows a very different pattern from that observed in temperate forests (Fig. 2.16(b)). At Lake Barrine, one group of species

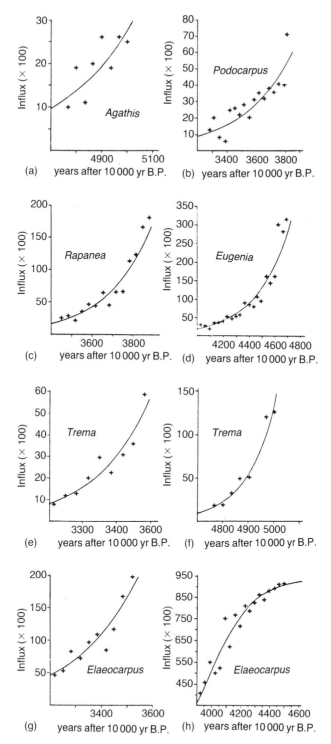

Figure 2.19 Pollen accumulation rates (PAR or pollen influx) reflecting population increases of representative rainforest tree taxa during the time interval of their invasion at Lake Barrine, Australia. Exponential curves are fitted to the PAR values in plots (a)–(g), and a logistic curve is presented in (h). From Walker and Chen (1987); reprinted with permission from Pergamon Press.

immigrated immediately after the change in fire regime, although the populations of the species increased individualistically, at different rates (Fig. 2.19), with population doubling times ranging from 100 to 300 years, on average much longer than those for temperate trees. For emergent, tropical rainforest species, such as dipterocarps, at least 60 years are required from germination to flowering (Webb and Tracey, 1981).

The second stage of species invasion and expansion at Lake Barrine occurred after 6000 years ago (Fig. 2.16(b)), followed by a third 'wave' of invasion after 5200 years ago. The continued immigrations of rainforest species represented the spread of species into what was already becoming a species-rich tropical forest. The specific sequence of invasion at Lake Barrine may have been related to the effectiveness of dispersal of the tree species, the number and degree of isolation of coastal refuge areas, and the distance across which the species had to move. For example, *Elaeocarpus* has fruits that are dispersed by birds (Webb and Tracey, 1981). One species of *Elaeocarpus* expanded with the first group of immigrants (Fig. 2.19(g)), and a second species invaded beginning 1800 years later (Fig. 2.19(h)), possibly due to chance dispersal from populations located initially farther from Lake Barrine. *Agathis*, a conifer with a relatively slow doubling time (Figs 2.16(b) and 2.19(a)), was among the third group of invaders; this tree may have preferentially colonized soils derived from basaltic and acidic volcanic rocks, geographically restricted edaphic sites where trees of *Agathis* persist today near Lake Barrine (Webb and Tracey, 1981). Thus, major controls on the Holocene sequence and pattern of establishment of rainforest tree species on the Atherton Tableland included climatically influenced disturbance regimes (wildfire), habitat availability, effectiveness of dispersal mechanism, and the distance and number of refuge areas from which the trees spread westward following climatic amelioration. No statistically significant trend through time towards an increase in the doubling times of tree populations is shown by the data from Lake Barrine (Figs 2.16(b) and 2.19). This means that once established, the tree populations provided no additional resistance to invasion by other species, which were continually being added to the assemblage. Therefore, there is no evidence that 'biological inertia' (Smith, 1965) was important in limiting the rates of invasion and population expansion of the trees that today comprise the tropical rainforest.

The Quaternary data from the tropics of Australia as well as from the Amazon Basin of South America (Colinvaux, 1987) show clearly that, during the last glacial/interglacial cycle, the tropical rainforest has dynamically expanded in area only in the past 6000 to 10 000 years. The formerly widely-held notion of ecologists that the tropical rainforest is a vegetation type characterized by long-term stability in species composition, set within an unchanging environmental disturbance regime and climate through evolutionary time, is untenable (Kershaw, 1981; Webb and Tracey, 1981; Hubbell and Foster, 1986; Colinvaux, 1987; Walker and Chen, 1987).

2.5 MIGRATIONAL STRATEGIES

The process of migration involves not only the dispersal of propagules but the successful establishment of reproducing populations as they extend the limits of distribution of a taxon beyond its previous boundaries (Delcourt and Delcourt, 1987a). Contour maps depicting changes in the dominance of plant taxa through time illustrate that rarely does the center of dominance of a taxon correspond with the geographical center of its total distributional range. Rather, the center of dominance is displaced geographically as a function of the migrational trajectory of its populations. In addition, the migrational responses of taxa are individualistic, with changes not only on the actively invading, leading edges of distribution, but also with progressive fragmentation and local extinction of populations occurring on retreating portions of range margins. As certain margins stabilize, expand beyond, or retreat from former range limits, changes occur in the overall shape, perimeter length, and degree of convolution of the perimeter of a taxon's distributional range. Some taxa, rather than migrating entirely from one geographic area to another, simply expand or shrink from one area to occupy a much larger or smaller area through time (Delcourt and Delcourt, 1987a).

Mapped patterns of changes in distribution can be analyzed in terms of different migrational strategies of taxa that are analogous to life-history strategies that determine the roles of taxa in plant succession (Delcourt and Delcourt, 1987a). In this analysis both the shape of the 'migration front' of invasion and the shape of the overall population changes behind the migration front are used to interpret the 'migrational strategy' of the taxon.

The 'r-strategist' is an early-successional taxon with the following characteristics (MacArthur and Wilson, 1967; Harper, 1977; Grime, 1979): it is relatively short-lived, has a high reproductive rate and wide dispersal of propagules, and it is shade-intolerant, germinating and growing in nutrient-poor, frequently disturbed habitats. By analogy, during a time of environmental change, 'r-migration (r_m) strategists' would tend to spread rapidly with a steep migration front and with declining percentage values of dominance (depicted by more widely spaced contour values in map view) behind the migration front (Fig. 2.20). Spruce (*Picea*) is an example of an r_m strategist (Delcourt and Delcourt, 1987a). During the late-glacial interval, with initial retreat of the Laurentide Ice Sheet, spruce advanced rapidly onto newly deglaciated terrain, with average rates of spread reaching 165 meters per year between 12 000 and 10 000 years ago. The northward advance of spruce was limited only by the physical barrier of the continental ice sheet (Bernabo and Webb, 1977; Ritchie and MacDonald, 1986). Behind the steep migration front, spruce populations diminished rapidly as climates warmed and eliminated spruce from forests to the south (Delcourt and Delcourt, 1987a).

The 'K-strategist' is contrasted with the r-strategist in being a late-successional taxon that is shade-tolerant and that allocates more of its

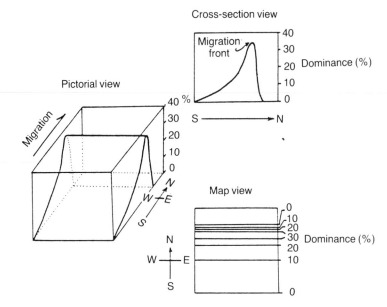

Figure 2.20 Hypothetical model of a plant population exhibiting an r-migration (r_m) strategy during a time of environmental change. From Delcourt and Delcourt (1987a).

resources to accumulating biomass, producing fewer but larger propagules that persist within the seed bank and whose dispersal is dependent upon gravity or animal vectors; K-strategists tend to grow on mesic, fertile sites (MacArthur and Wilson, 1967; Harper, 1977; Grime, 1979). On a Quaternary time scale, the 'K-migration (K_m) strategist' can be depicted (Fig. 2.21) by a cross-sectional view of percent dominance that is sigmoid-shaped, reflecting a more gradual rise to dominance after initial invasion than is exhibited by the r_m strategist. Highest values of percent dominance may occur a large distance away from the migration front, reflecting a slow build-up of populations over time. Oak (*Quercus*) is an example of a K_m strategist. Across eastern North America, oak was a minor constituent of late-glacial forests. Although the northern limits of distribution of oak reached to the Great Lakes region during late-glacial times, as a genus, oak increased to high values of percent dominance only during the Holocene in the region today occupied by eastern deciduous forest (Delcourt and Delcourt, 1987a).

A third migrational strategy is associated with fugitive species, whose local populations are typically restricted to special habitats in which competition from other taxa is limited or which are ephemeral in time because of repeated disturbances (Harper, 1977; Grime, 1979). The 'fugitive-migration (f_m) strategist' (Fig. 2.22) would occupy a patchwork array of geographically isolated sites (Delcourt and Delcourt, 1987a). Through glacial–interglacial

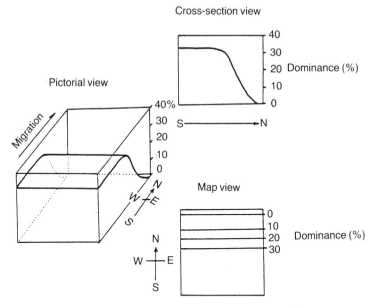

Figure 2.21 Hypothetical model of a plant population exhibiting a K-migration (K_m) strategy during a time of environmental change. From Delcourt and Delcourt (1987a).

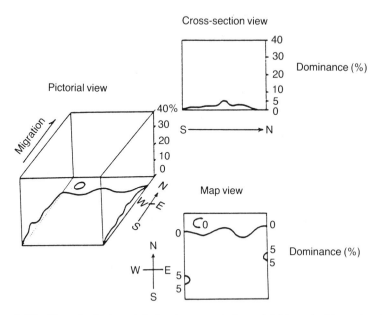

Figure 2.22 Hypothetical model of a plant population exhibiting a fugitive-migration (f_m) strategy during a time of environmental change. From Delcourt and Delcourt (1987a).

cycles, fugitive populations might exhibit a temporally ephemeral 'archipelago' of suitable islands of habitat, shifting geographically in response to changes in disturbance regimes, physical characteristics of landscape patches, and availability of sites in which competition from other species is minimized. On a time series of changing population maps for the f_m strategist tamarack (*Larix laricina*), the map patterns reflect this species' occupation of wetland environments that change in distribution through time. The generalized distributional pattern of tamarack at any one time thus has low dominance values over most of its range with high dominance values only in local wetland areas (Delcourt and Delcourt, 1987a).

2.6 CONCLUSIONS

1. Paleoecological studies have verified and extended historic observations on the processes of dispersal, invasion, and population expansion that are the outcome of life-history characteristics of species.

2. In general, effective dispersal of seeds has not been a limiting factor to postglacial invasions of temperate trees. Potential animal dispersers of nut-bearing temperate trees have been identified for which fossil evidence is available, indicating that in most cases heavy-seeded, mast-bearing trees did not lag behind the migrations of light-seeded, wind-dispersed trees on account of the lack of an effective dispersal mechanism. However, in certain circumstances, postglacial invasions of both tropical and temperate trees which have large, fleshy or dry fruits may have been significantly impeded by loss of late-Pleistocene large mammals that were formerly important seed dispersers.

3. The rates of Holocene population expansions of both temperate and tropical trees appear to be related more to the individualistic life-history characteristics of the species than to biological inertia of the communities into which they invade, even for species-rich tropical rainforest.

4. Models for rate of range expansion based on the observed behavior of recently introduced species can be tested using the empirical fossil record of range expansions of species associated with the shift from glacial to interglacial conditions. This comparison reveals that both biological and climatic controls have been important in determining the rates of range expansion for temperate trees during the Holocene. However, the timing for changeover from biological control and exponential expansion of populations to climatic control and reduced rates of spread differs among taxa.

5. Through Quaternary map analysis, plant taxa are demonstrated to exhibit migrational strategies that are analogous to the role that life-history strategies play in plant succession.

3 Plant succession

3.1 ISSUES

Concepts of plant succession have had a long history of development in both European and North American ecology (McIntosh, 1985). Turn-of-the-century ecologists including Cowles (1899), Clements (1916) and Shelford (1911, 1913) observed the modern patterns of species distributions on sand dunes and around the margins of ponds and in bogs. From these observations of vegetation zonation in space came interpretations of the history of successional change in community composition through time. Specifically, the concept of primary succession was developed to account for changes in vegetation that occurred from the time of colonization of bare dune sand to the development of mature hardwoods forest on stabilized dunes (Cowles, 1899; Olson, 1958).

The assumption that a spatial series of extant plant communities serves as an adequate representation of a true chronological series of developmental changes has been accepted by ecologists for decades (Drury and Nisbet, 1973) and still forms much of the basis for generalizations about plant succession (West et al., 1981). However, this assumption of space-for-time substitution has been questioned in recent years (Walker, 1970; Pickett, 1989), primarily by plant community and population ecologists who favor a more reductionist approach to plant succession based upon population changes (Peet and Christensen, 1980) that may involve stochastic responses which vary with life-history characteristics and physiological tolerances of available species (Shugart, 1984). Despite the availability of paleoecological means of investigation throughout the history of modern ecology (von Post, 1916), few studies of long-term vegetational dynamics have been conducted explicitly to test the critical assumption that changes in plant communities observed across space adequately reflect compositional changes that have actually occurred through time (Pickett, 1989).

Many studies of plant succession focus on secondary succession, that is, changes in plant communities that have occurred after disturbance (White, 1979; Pickett and White, 1985). Further, studies of secondary succession in North America have tended to concentrate on the regeneration of secondary temperate forest after abandonment of agricultural old-fields (Odum, 1969).

Although the trends observed may be characteristic of large regions today, their usefulness in making universal generalizations in ecology may be limited to the extent that the observations are based upon systems severely altered by anthropogenic activity, and therefore unrepresentative of natural processes that have occurred prior to major human impacts. Because of the current worldwide extent of human modification of ecosystems, paleoecological techniques are now the major source of information concerning plant succession under natural conditions.

Plant succession in temperate forested regions takes place through replacement of tree populations over a time span of several hundred years to a thousand years (Olson, 1958; Loucks, 1970; Peet and Christensen, 1980; Oliver, 1981; Shugart, 1984). One reason that paleoecological data have been ignored in evaluation of the nature of plant succession is because of the widely held and erroneous assumption that these data can only be obtained on a scale of resolution that makes them applicable to questions on an evolutionary time scale rather than a shorter-term successional time scale (Drury and Nisbet, 1973; Pickett, 1976). However, recent studies now demonstrate that paleoecological data are sensitive recorders of vegetational changes over the past 500 years (Brugam, 1978a,b; Bradshaw and Miller, 1988; Jackson *et al.*, 1988).

Loucks (1970; Fig. 3.1) emphasized the importance of understanding the influence of disturbances that 'reset' vegetation to early stages of succession. Various forms of disturbance are today recognized as important agents of change in plant communities. Disturbance agents such as wildfire are important in maintaining mosaics of communities in a variety of successional states across broad landscapes (Bormann and Likens, 1979; Romme and Knight, 1982; Pickett and White, 1985). Debris avalanches in mountainous terrains and cryoturbation in permafrost environments are types of geomorphic disturbances that affect pattern and process in vegetational devel-

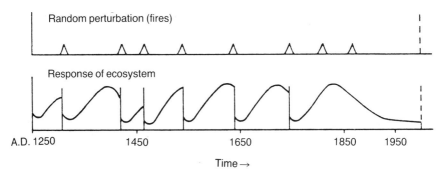

Figure 3.1 Ecosystem response to random disturbances that restart plant succession and accompanying changes in biomass, primary productivity, diversity of seedlings and other ecosystem attributes in temperate forests. Modified from Bormann and Likens (1979).

opment (Drury and Nisbet, 1973; Pickett and White, 1985; Swanson *et al.*, 1988). The effects of these agents of disturbance can be studied through analysis of the paleoecological record.

Observations of vegetational change over many episodes of plant succession are important to understanding whether the observed trends are cyclic and predictable, or whether they are strictly the outcome of random events (Horn, 1976; Shugart, 1984). Paleoecological data spanning an interval of as little as 500 to several thousand years can provide data pertinent to resolving questions of successional responses to environmental disturbances, as well as establishing recurrence intervals for disturbance regimes such as wildfire (Green, 1981; Clark, 1988a, b, c, 1990). A further question is whether plant successional trends are governed entirely by changes in the physical environment (allogenic succession), or whether the plant community itself plays a primary role in governing the direction of successional change through time (autogenic succession; Tansley, 1935). By examining vegetational changes occurring during times of relative constancy in climate and in situations where disturbance is minimal, paleoecological data can evaluate this question of environmental or biological control over plant succession (Payette, 1988).

Many models of succession, even ones predicated on evolutionary responses of populations (Pickett, 1976), assume either constant environments or gradual environmental changes through time (Loucks, 1970; Bormann and Likens, 1979; Pickett, 1976). These conceptual models depict successional changes occurring over many cycles to result in a dynamic equilibrium (Figs 3.1 and 3.2) in which climate, species availability, and disturbance regimes

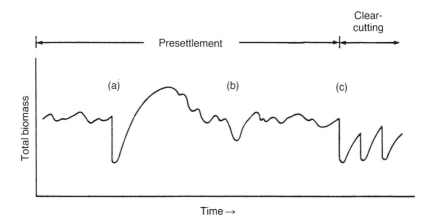

Figure 3.2 Hypothetical pattern of changes in total biomass in temperate forests of eastern North America during presettlement times when infrequent disturbances included (a) fire and (b) wind, and (c) during postsettlement times, with harvesting of forests occurring at regular intervals of about 60 years. Modified from Bormann and Likens (1979).

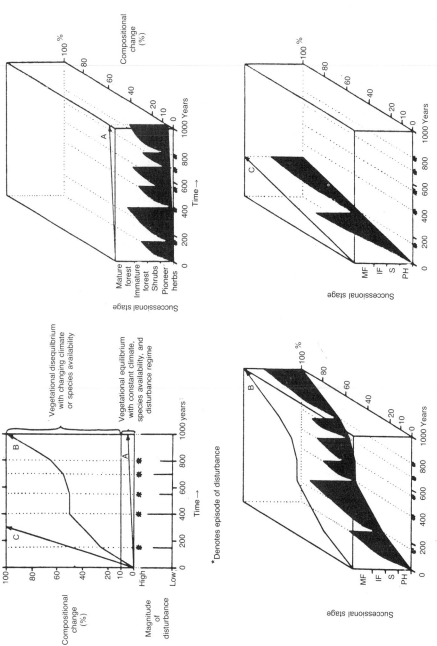

Figure 3.3 Trajectories of change in composition and structure of vegetation as influenced by changes in both climatic and disturbance regimes. Trajectory A represents vegetational equilibrium reflecting stable climate, species availability and disturbance regime, with the saw-toothed curve depicting changes in biomass following disturbances that restart plant succession. Trajectory B represents vegetational disequilibrium as the composition of vegetation changes dramatically in response to 400 years of climatic change followed by 200 years of stable climate, and then a return to changing climate over the next 400 years. Trajectory C illustrates dynamic vegetational disequilibrium accompanying a rapid climatic change of great magnitude. From Delcourt *et al.* (1983); reprinted with permission from Pergamon Press.

are essentially constant and for which the degree to which succession proceeds is largely a function of frequency and intensity of random disturbances such as windthrow or wildfire (Loucks, 1970; Pickett, 1976; Bormann and Likens, 1979). This interpretation is similar to the hypothetical scheme presented in trajectory A in Fig. 3.3 (Delcourt *et al.*, 1983). Such an interpretation may be misleading, however; over even the last 1000 years of the late Quaternary interval, climatic changes of the magnitude of the Little Ice Age (circa A.D. 1640 to 1880, Fig. 3.4(a)) have greatly influenced fire regimes and the sediment yield eroded from watersheds, represented by charcoal accumulation (Fig. 3.4(b)) and thickness of lake-sediment varves (Fig. 3.4(c); Clark, 1988c). Plant-successional response to more rapid environmental changes (trajectory B, Fig. 3.3) on this time scale would be expected to be characterized by vegetational disequilibrium, with the composition of late-successional communities changing through time (Delcourt *et al.*, 1983). In a more extreme case of very rapid climatic change, such as that potentially induced over the next two centuries by global climatic warming due to anthropogenic additions of atmospheric CO_2 and other greenhouse gases (Schneider, 1989), it is conceivable that a complete turnover in species composition could occur within only two plant-successional

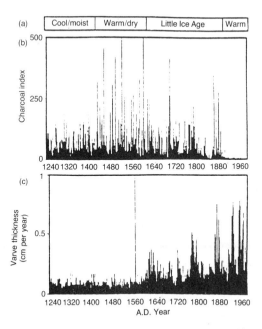

Figure 3.4 (a) Changes in climate of the northern Great Plains since A.D. 1240 as reflected in (b) changes in accumulation rates of charcoal and (c) changes in thickness of varved sediments from Deming Lake, Minnesota. Modified from Clark (1988c).

cycles (trajectory C in Fig. 3.3; Delcourt *et al.*, 1983). On this scale, environmental changes will dominate successional responses (Delcourt *et al.*, 1983) and may even override interactions of evolutionary strategies of constituent species that otherwise might be responsible for observed successional trends (Pickett, 1976).

3.2 PRIMARY SUCCESSION ON GLACIAL MORAINES

Are there consistent, predictable patterns in structural change, species richness and functional attributes such as nutrient cycling that characterize plant succession through time? This is one of the central issues debated in the literature on plant succession that has led to adoption of a reductionist view by many ecologists (Peet and Christensen, 1980). Odum (1969, reiterated by Bormann and Likens, 1979) proposed that both the structure and functioning of ecosystems change in predictable ways through the course of succession. These include increases in biomass (Fig. 3.2), in species diversity, and in efficiency of nutrient cycling (Bormann and Likens, 1979). The predictability of successional change has been challenged through citation of counter-examples based on empirical studies that show opposite or ambiguous trends (Drury and Nisbet, 1973) as well as through the use of Markovian chain models (Horn, 1976) that indicate the importance of both random events and population dynamics of individual species in determining the outcome of plant succession (Shugart, 1984).

We have an opportunity to examine these questions concerning space-for-time substitution from a paleoecological point of view. Are the spatial trends in primary plant succession observed on modern landscapes directly analogous to the chronological sequences of successional changes that are documented from paleoecological evidence from sites in the same region?

3.2.1 Modern patterns

In a series of papers, Herb Wright and colleagues documented a number of aspects of modern deglaciated environments in the vicinity of the currently retreating Klutlan glacier in the St. Elias Mountains of the western Yukon Territory of Canada (Wright, 1980). Mapping of glacial geology provided the environmental setting for development of lakes on moraines of differing ages. The vegetation and soils surrounding the lakes were documented, and relationships of modern pollen rain to quantitative vegetation composition were established. The successional relationships inferred from the vegetational sequence occupying successively older moraines (with land surfaces formed as long as 1200 years ago) was used as a model for interpreting the changes in composition of fossil pollen assemblages from several lakes located on recessional moraines as well as in the uplands. This research design was intended to test whether modern successional relationships were suitable analogues for the developmental history recorded in individual lakes to allow extrapolation to plant succession occurring in the midwestern United States at the end of the Pleistocene (Wright, 1980).

On a series of five successively older moraines of the Klutlan glacier, plant succession today begins with an open gravel substrate dominated by *Crepis nana*, *Dryas drummondii* and *Hedysarum mackenzii*, a pioneer community including nitrogen fixers and forming dense mats that trap wind-blown silt on which acrocarpous mosses establish. After about 75 years a layer of sedimentary debris forms as a cover over stagnant blocks of glacial ice, allowing seedlings of willow (*Salix*) and *Shepherdia canadensis* to establish on the land surface. The shrubs provide shade for pleurocarpous mosses, and white spruce (*Picea glauca*) seedlings are first found rooted in the moss carpet underneath 40-year old shrubs. *Shepherdia* is gradually replaced by a *Picea–Salix* woodland with *Arctostaphylos uva-ursi* shrubs; as the *Picea* trees mature, a dense spruce forest eventually dominates (H. J. B. Birks, 1980a).

That this spatial pattern of plant communities reflects a true temporal sequence of plant invasion was confirmed by analysis of plant macrofossils obtained at intervals in the organic soils beneath the older spruce forest; leaves of *Dryas* were found at the base of the soil, overlain by remains of shrubs, with needles of *Picea* in the surface layers (Wright, 1980). Pollen analysis of Triangle Lake on the Harris Creek moraine showed a series of paleoecological changes through time that were similar to the spatial pattern observed on the series of recessional moraines (H. J. B. Birks, 1980b). At Triangle Lake, *Shepherdia/Salix* shrub vegetation was replaced by open *Picea* woodland and then by closed *Picea* forest (H. J. B. Birks, 1980b). Pollen analysis of lake sediments from Gull Lake, located in open spruce forest in the uplands adjacent to the newly-formed glacial moraines, also revealed this general successional sequence. The major difference was in the vegetation that first established after Gull Lake formed. At the base of the lake sediments was a layer of volcanic ash that was deposited 1220 radiocarbon years ago. The plant-fossil assemblage consisted primarily of spruce and sedge, interpreted as an herb tundra colonizing on the volcanic ashfall, and representing a community different from any on the modern transect. Shrub tundra then developed on the uplands surrounding Gull Lake, with increases in pollen of birch (*Betula*), *Alnus crispa*, and *Salix*, with open-ground environments indicated by *Artemisia*, *Dryas*, sedges, and grasses. Open *Picea–Betula* woodland developed by about 800 years ago (H. J. B. Birks, 1980b; Wright, 1980).

3.2.2 Late-glacial environments

Application of the results from the study of the Klutlan glacier to the late-glacial of Minnesota requires appropriate temporal resolution to be obtained from the late Pleistocene sites in the midwestern United States. Thus, the first 500 years of vegetation history are critical to understanding the process of initial colonization of the newly available landscape. The majority of Minnesota lakes originated as shallow depressions in stagnant ice blocks left behind buried in glacial till. As the blocks of ice gradually melted, 'kettle' lake basins formed. The first sediments deposited in these lakes often represent the vegetation that was growing on top of glacial debris that initially covered the ice blocks, and complete melting of the ice has been shown to lag behind the regional deglaciation by as much as several thousand years (Wright, 1980).

Paleoecological evidence from a number of lake sites in central Minnesota confirms that the landscape was fully vegetated at the time of initiation of the kettle lake depressions, analogous to the situation in the Yukon where vegetation established within a few decades of stabilization of the till mantle over stagnant ice. In central Minnesota, some sites like Wolf Creek and Kylen Lake record evidence for treeless communities of 'tundra barrens' persisting from full-glacial through late-glacial time in the periglacial margin at the southern edge of the Laurentide Ice Sheet (Birks, 1976, 1981a). However, most late-glacial sites across the Great Lakes region lack evidence for true arctic tundra at the time of lake formation, probably reflecting the time lag in formation of kettle lakes after climatic warming and postglacial retreat of the continental ice sheet. Typically found in late-glacial lake sediments, plant macrofossils in the initial forest floor 'trash layer' include pioneer plants common to both tundra and boreal forests, such as *Dryas integrifolia* and *Arctostaphylos uva-ursi*. The corresponding fossil remains of the beetle fauna are also generally indicative of forest/tundra transition with forest openings. Following the glacial retreat, the pioneer colonization phase may have lasted only a few decades and is characterized by open-ground communities represented by pollen of sedges and *Artemisia* (Ashworth *et al.*, 1981). Establishment of tree populations may have taken longer in the Minnesota cases than in the modern Klutlan situation, because of a time lag introduced by the greater distance from which *Picea* had to migrate to reach the central Minnesota region during the late-glacial interval (Wright, 1980).

The comparison of recent successional trends at the Klutlan glacier with those that occurred some 13 000 years ago in Minnesota illustrates that the modern sequence bears certain similarities to the ancient primary successional patterns (Ashworth *et al.*, 1981). In particular, the transition from pioneer herbaceous communities to forest communities proceeded through a phase in which nitrogen-fixing herbs and shrubs were important functional members of the community. However, this comparative study also demonstrates that the specific course of plant succession was in large part determined by initial environmental conditions and by subsequent species availability. In the Klutlan example, at Gull Lake the initial substrate was blanketed by a layer of volcanic ash that was first colonized 1220 years ago by an assemblage of plants with no counterpart either in the modern vegetation of the Klutlan moraines or in the late-glacial tills of Minnesota. As demonstrated at Triangle Lake, in the Yukon Territory, spruce trees have been present during the last millennium for rapid colonization; in late-glacial Minnesota, primary succession occurring during the first 500 to 1000 years after deglaciation took place without a local seed source for spruce trees. Finally, at Klutlan today obligate tundra plants exist and are important pioneer plants; in late-glacial Minnesota many true tundra plants had already disappeared from the region before many of the lakes stabilized and began to accumulate a sediment record.

3.3 HYDROSERE SUCCESSION

Hydrosere succession is traditionally considered to be a form of primary succession that ultimately results in changes from hydric to mesic communities. Hydrosere succession was first postulated in the early twentieth century by Shelford (1911, 1913) and Clements (1916) to account for the present-day zonation of aquatic plant communities at the margins of lakes and bogs. A critical assumption of the hydrosere succession hypothesis is that autogenic processes control changes in aquatic vegetation through time, and that with the gradual filling in of a lake basin with organic remains of plants, the stages in succession are predictable, with assemblages of submersed aquatic macrophytes replaced in sequence by floating-leaved, rooted aquatics, then by emergent plants, and with ultimate convergence of plant communities on mesic forests (Walker, 1970; Jackson *et al.*, 1988). The seral stages observed on the modern landscape, however, have been shown by paleoecological studies to give a false impression not only of the predictability of successional changes through time but also of the relative influence of autogenic and allogenic controls on plant succession (Walker, 1970; Jackson *et al.*, 1988).

3.3.1 Multiple pathways for autogenic hydrosere succession

In a comparison of the developmental history of a large number of wetlands in Great Britain, studied through the use of stratigraphic and plant-fossil studies of both lake and peat-bog sediments, Walker (1970) demonstrated that, even under situations controlled primarily by autogenic factors, a variety of successional pathways is possible, and hydrosere succession does not inevitably culminate in mesic woodland communities. Walker (1970) found that the changes in plant-macrofossil assemblages in wetland sites through the Holocene interval were generally unrelated to episodes of climatic change, but rather that they reflected local, autogenic vegetational development. With the paleoecological record, he tested the hypothesis of hydrosere succession of Tansley (1939), which was adapted for British vegetation from ideas initially proposed by Shelford (1911, 1913) and Clements (1916) in North America.

Tansley (1939) proposed that initial colonization of aquatic macrophytes in open water was followed by reeds and bulrushes. With accumulation of decaying organic remains produced by the reed swamp, the water level would become more shallow, and marsh and fen plants would become established. With further raising of the soil surface by accumulation of organic matter, shrubs such as alder would replace fen with a scrub or woodland vegetation locally known as 'carr'. Eventually the soil surface would no longer be waterlogged, and mesic forest would replace the shrub carr. Evidence for ultimate succession to mesic forest was circumstantial and based upon observations of oak forest growing at the margins of certain tracts of carr vegetation (Walker, 1970). Succession from fen to *Sphagnum* bog was recognized as a special case that was postulated to occur primarily in areas of moist oceanic climate where high precipitation contributed to the success of *Sphagnum* moss in expanding across landscapes and building raised bogs (Tansley, 1939).

Walker (1970) analyzed 20 published plant-fossil records from throughout the British Isles for evidence of hydrosere succession. He found that the predominant sequence of succession was from microorganisms or submersed macrophytes to floating-leaved macrophytes, to reed swamp, to fen, to fen carr, to bog; this sequence, however, occurred in only 46% of observed cases. In the remainder of the sequences, alternative successional trends were observed. In particular, the timing of establishment of *Sphagnum* was a critical factor determining the future course of a given successional sequence because of its ability to change the pH of the site as well as to sustain its own growth independently of lake water table. Walker (1970) concluded that, overall, community types recognizable today, such as reed swamp and fen, can be recognized in the fossil record, and their development has been related temporally to one another at a number of locations. However, given the variety of observed alternate successional pathways, specific predictions about the future course of succession in a given wetland can only be made in terms of probability of occurrence, even in situations free from allogenic influences.

Through the course of postglacial succession in British wetlands there has been a general tendency toward progressive reduction of depth of open water in aquatic stages of succession, with an increased dependence on direct rainfall to sustain the terrestrial communities. In this respect the observed vegetation zones surrounding lakes are analogous in a general way to the sequence of vegetation types that may develop over time. However, the range of specific vegetation types present at a site at one time may or may not include all the past plant communities that have actually existed during its development. Nor will they necessarily be useful in predicting the specific successional sequence that will occur in the future, except in terms of probabilities determined from considering all possible transitions from the modern vegetation in the context of the environmental conditions at the site and the availability of species. Overall, paleoecological records from across the British Isles document that, rather than mesic forest, the predominant natural 'climax' vegetation resulting from hydrosere succession is *Sphagnum*-dominated bog (Walker, 1970).

3.3.2 Influence of historical factors on hydrosere succession

The major concepts of hydrosere succession were originally proposed for the Great Lakes region by Shelford (1911, 1913) and Clements (1916). Cowles (1899) first recognized that the sand dune complexes adjacent to southeastern Lake Michigan represented a temporal sequence, with oldest dune ridges located farthest inland from the modern shoreline. Plant communities, both in the uplands and within interdunal ponds (Fig. 3.5), were thought to be arrayed in a chronosequence whose modern composition was directly related to length of time since dune and pond formation (Cowles, 1899; Shelford, 1911, 1913). Youngest ponds (rows 1 and 2 in Fig. 3.5) are dominated by

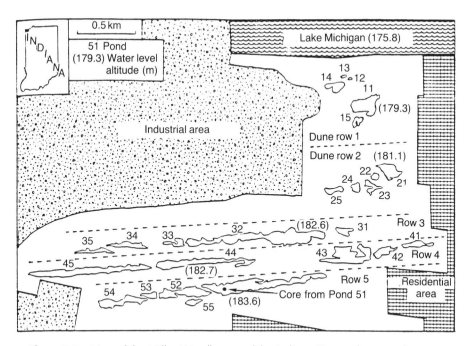

Figure 3.5 Map of the Miller Woods area of the Indiana Dunes along southeastern Lake Michigan showing the location of the sediment core from interdune Pond 51. Modified from Jackson *et al.* (1988).

submersed aquatic macrophytes including *Chara, Najas flexilis, Potamogeton, Myriophyllum*, and *Utricularia vulgaris*. Ponds in rows 3–5 (Fig. 3.5) are arrayed along a gradient from vegetation dominated by submersed and floating-leaved macrophytes (the latter including *Nuphar* and *Nymphaea tuberosa*), with the presence of some emergents (*Proserpinaca palustris, Scirpus acutus, Sparganium chlorocarpum*, and *Typha angustifolia*), to oldest ponds dominated by emergents as well as shrubs of *Cephalanthus occidentalis*. Originally postulated by Shelford (1911, 1913), the hypothesized successional sequence started with an open beach pool surrounded by bare dune sand substrate that was invaded by a mixture of aquatic vegetation including submersed, floating-leaved, and emergent life forms. Through time, the aquatic vegetation became dominated by emergent plants and finally shifted to shrub-dominated vegetation. Shelford's idealized sequence of hydrosere succession was interpreted as the long-term product of autogenic processes, predominantly reflecting the accumulation of organic-rich sediments and lowering of relative water levels in the ponds.

Analysis of pollen and plant macrofossils from a sediment core from one of the oldest ponds (Pond 51; Figs 3.5 and 3.6) allowed major reinterpretation of the sequence of vegetational change through time in the Indiana Dunes along Lake Michigan and challenges the concept of hydrosere succession as originally proposed (Jackson *et al.*, 1988). Pond 51 formed 2700 years ago, during the interval of diminishing lake levels from a previous mid-Holocene high stand (transition from Nipissing Stage to Algoma Stage) of Lake Michigan. The plant-macrofossil data showed no trend toward changes in species composition from 2700 years ago to about 150 years ago (40 cm depth). The time of regional land clearance by EuroAmerican pioneers was marked in the Pond 51 sediment record by an increase in pollen of ragweed (*Ambrosia* type) and a decline in pine pollen. Before the time of EuroAmerican settlement, the vegetation of Pond 51 was a diverse assemblage consisting of submersed (*Najas, Chara, Nitella*), floating-leaved (*Brasenia, Nuphar*), and emergent macrophytes (*Cyperus, Eleocharis, Polygonum, Bidens, Leersia, Scirpus, Dulichium, Eupatorium, Zizania* and *Carex*). However, the plant-fossil assemblage that accumulated in sediments dating after 150 years ago showed major changes in composition, primarily with a conspicuous increase in pollen of *Typha* and *Cephalanthus* and decreases in macrofossil representation of numerous submersed and floating-leaved macrophytes.

Jackson *et al.* (1988) concluded that the modern composition of aquatic vegetation, and even that at the time of study of Shelford, has been affected to a great extent by allogenic factors including historic human disturbance. In the Millers Woods tract in which Pond 51 is located, a railroad grade was in place adjacent to the site as early as 1851. Small particles of clinkers produced by nearby industrial activity are present in post-settlement sediments. Human activities caused changes in local hydrology as well as in sedimentation to the dune ponds, distorting the natural conditions and altering the vegetation

MILLER WOODS POND 51, Core A

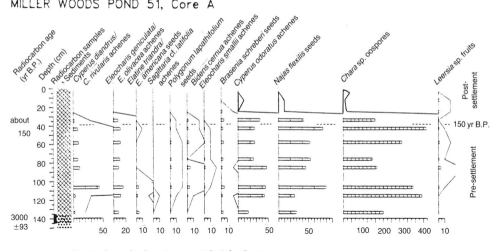

Number of macrofossil specimens per 115 ml of sediment

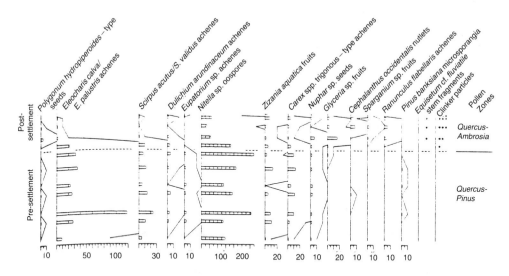

Number of macrofossil specimens per 115 ml of sediment

Figure 3.6 Plant macrofossil diagram from Miller Woods Pond 51 showing changes through the past 3000 years in the number of fossil specimens of shrubs, as well as of emergent, floating-leaved and submersed aquatic plants contained in the sediment samples from Core A. Modified from Jackson *et al.* (1988).

from its presettlement condition. Jackson *et al.* (1988) concluded that the differences observed today in vegetation among ponds along the Indiana Dunes chronosequence result from differences in type and magnitude of disturbance events rather than from differences in autogenic successional stages. Although the overall sequence of vegetational history at Pond 51 bears similarities to the developmental sequence postulated by early ecologists, the specific compositional changes through time differed from those observed today. The driving mechanisms for observed successional changes were not recognized by the previous, long-held hypothesis of chronosequences proposed by Cowles, Shelford and Clements.

3.4 UNIDIRECTIONAL SUCCESSIONAL CHANGES

Another aspect of succession that can be approached using paleoecological techniques involves the question of whether vegetational changes accompanying gradual environmental change, for example, sea-level rise relative to the land surface, are unidirectional and thus predictable, or whether a series of alternate vegetational states is possible depending upon the local patchwork of the landscape mosaic.

3.4.1 Successional responses to sea-level rise

Alfred Redfield (1965, 1972) described the development of Barnstable Marsh, Massachusetts, as determined by the interaction of four major factors: (1) the tidal range; (2) the physiological tolerances of peat-producing marsh plants to changing tide levels; (3) sedimentation processes, both on open tidal flats and within stands of vegetation; and (4) changes in sea level relative to the land. Redfield confined his study to salt marsh vegetation, particularly that dominated by *Spartina alterniflora*. In Redfield's successional scheme (Fig. 3.7), intertidal marsh is transformed into high marsh gradually over hundreds of years through vertical accretion of sediment and development of increasingly dense stands of *Spartina*. Once the surface of the peat reaches

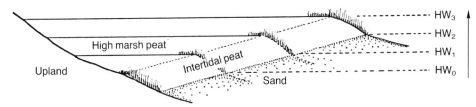

Figure 3.7 Theoretical environmental changes in coastal New England as the salt marsh spreads inland over sediment that accumulated on a tidal sand flat during an episode of rising sea level. From Redfield (1972).

high water level, continued salt marsh development over newly inundated land is dependent upon a progressive rise in sea level. At Barnstable Marsh, for example, Redfield (1972) documented over 6 meters of peat accumulation during the past 3600 years. In Redfield's study, plant succession was envisioned as a gradual process that involved successful colonization of newly exposed tidal flats by salt marsh grass, and the gradual accumulation of salt marsh peat in accord with the relatively slow changes in elevation and areal extent of marsh habitat resulting from gradual shifting of sea level through time. This view of salt marsh succession emphasized the predictability of changes in vegetation zones through time.

Jim Clark (1986) challenged the interpretations of Redfield as being oversimplified in the context of the complexity of both short-term and long-term environmental changes that affect coastal processes (Clark and Patterson, 1985; Clark, 1986). Clark (1986; Fig. 3.8) considered a wide range of habitats, from fresh water to salt water and from tidal inlets to barrier-beach uplands, in his successional scheme. He examined a series of sediment cores for both stratigraphy of sediment types indicating changes in physical environment and for fossil pollen content that gave a detailed reconstruction of changes in plant communities over a time frame of the past several hundred years. He found that the vegetation changed dynamically as a result of disturbance episodes that were primarily storm-driven. However, he found no systematic pattern of replacement of one plant community type by another (Fig. 3.8) because of the spatial and temporal complexity of the disturbance regime. Clark (1986) concluded that, inasmuch as tidal inlet activity influences vegetational change in submersed-aquatic, marsh, and fringe assemblages through adjustments in salinity and tidal range, species composition could be predicted by environmental conditions but not by successional stage.

3.4.2 Retrogressive changes in plant communities

Succession that leads to communities with increased complexity or biomass within a habitat that becomes more mesic with time may be thought of as progressing toward a late-successional community on mature, nutrient-rich soil. In contrast, 'retrogressive' changes resulting from soil degradation lead to either hydric or xeric communities that are poor in species and have reduced biomass (Barbour *et al.*, 1987). Evidence for retrogressive changes in plant communities has come from two primary sources: (1) chronosequences of vegetation such as those on California coastal terraces (Jenny *et al.*, 1969); and (2) fossil pollen studies of Holocene soil profiles in woodlands such as those of northwestern Europe (Iversen, 1969; Stockmarr, 1975).

The pygmy cypress (*Cupressus pygmaea*) forests of Pleistocene coastal terraces in Mendocino County, California, are an example of relatively open-stature, low biomass vegetation that persists on edaphically suitable

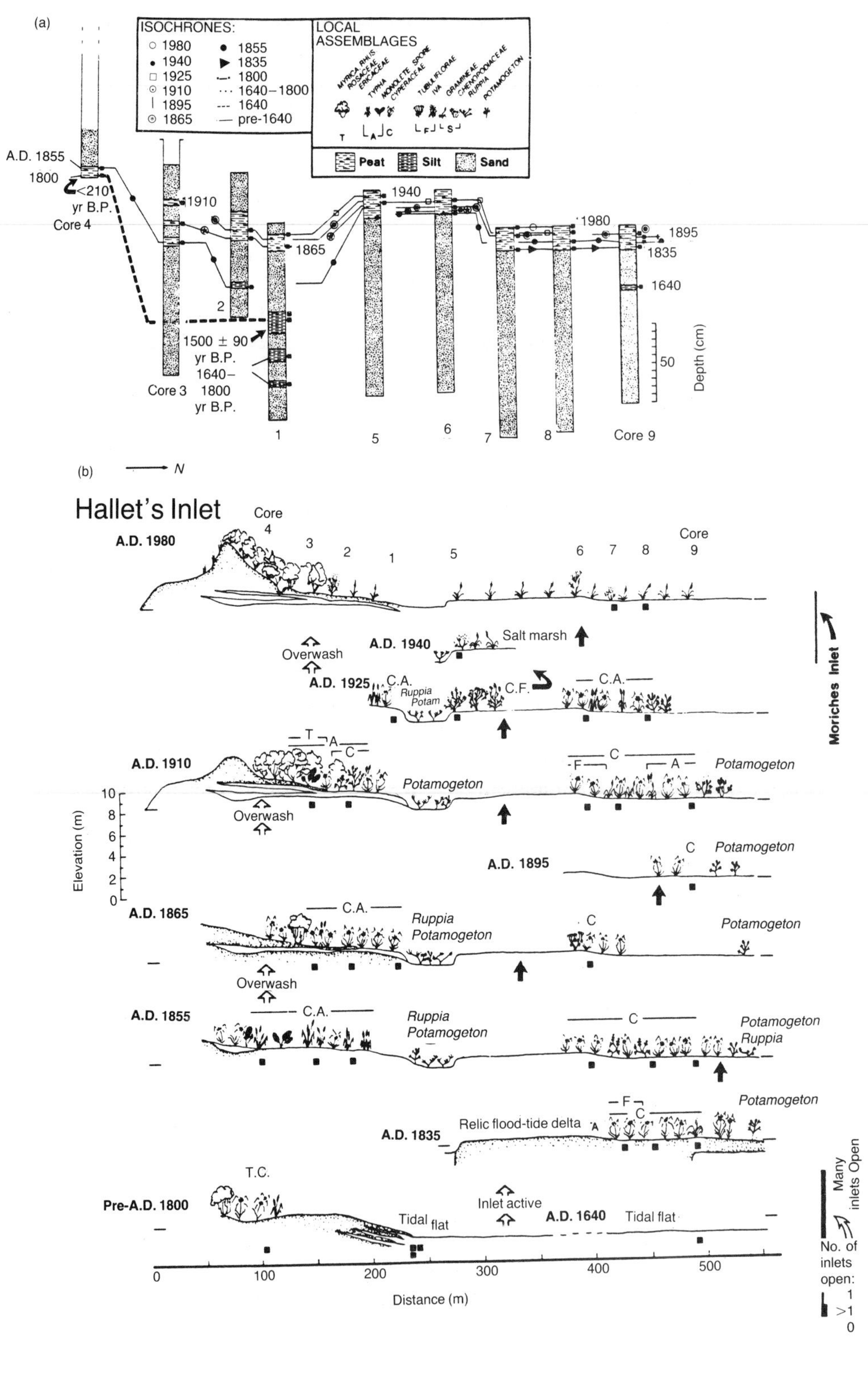

(a)

ISOCHRONES:
○ 1980   ● 1855
● 1940   ▶ 1835
□ 1925   --- 1800
⊙ 1910   ··· 1640–1800
| 1895   ---- 1640
⊙ 1865   — pre-1640

LOCAL ASSEMBLAGES

Peat   Silt   Sand

A.D. 1855
1800
<210
yr B.P.
Core 4

1910
1865
1940
1980
1895
1835
1640

2

1500 ± 90
yr B.P.
1640–
1800
yr B.P.

Core 3

Depth (cm)
50

1   5   6   7   8   Core 9

(b) → N

Hallet's Inlet

Core
A.D. 1980   4   3   2   1   5   6 7 8 Core 9

Overwash   A.D. 1940   Salt marsh ↑

A.D. 1925   C.A.   C.F.   C.A.
Ruppia
Potam

A.D. 1910   — T — A   — C —   Potamogeton
                 — C —   F — C — A
Overwash   Potamogeton

Elevation (m)
10
8
6
4
2
0

A.D. 1895   C   Potamogeton

A.D. 1865   — C.A. —   Ruppia   C   Potamogeton
Overwash   Potamogeton

A.D. 1855   — C.A. —   Ruppia   — C —   Potamogeton
Potamogeton   Ruppia

A.D. 1835   — F —   Potamogeton
Relic flood-tide delta   A   C

Pre-A.D. 1800   T.C.
Inlet active
Tidal flat   A.D. 1640   Tidal flat

0   100   200   300   400   500
Distance (m)

Moriches Inlet →

Many
inlets Open

No. of
inlets
open:
1
>1
0

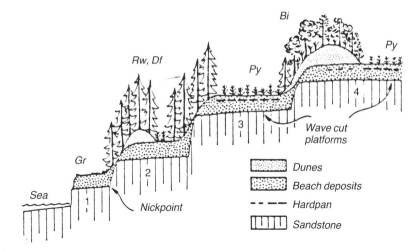

Figure 3.9 Schematic drawing of four marine terraces from the Mendocino Coast of California. Gr = Grassland; Rw, Df = Redwood–Douglas fir forest; Bi = Bishop pine forest; Py = Pygmy cypress forest. From Jenny *et al.* (1969).

sites within a region whose mesic, late-successional forests contain giant redwood (*Sequoia sempervirens*) and Douglas fir (*Pseudotsuga menziesii*). The pygmy cypress forests contain dwarf Ericaceous shrubs in addition to numerous endemic species and edaphic ecotypes. This vegetation type is confined to highly leached, acidic podzol soils with hardpans that promote local ponding of water. These soils occur geographically and topographically restricted to Pleistocene marine terraces that have been raised as much as 200 meters above today's mean sea level by recent tectonic uplift (Fig. 3.9), with the oldest terraces estimated to date at least several hundred thousand years before present. Evolution and continued persistence of the 'retrogressive' pygmy cypress forest was attributed by Jenny *et al.* (1969) to a combination of Quaternary geological and soil-forming processes leading to soil conditions that stunted plant growth, produced dwarfism, and promoted ecotypic differentiation as well as narrow endemism.

Direct evidence of the effects of changes in soil properties through time with consequent retrogressive effects on forest vegetation is provided by the studies of Iversen (1969) and Stockmarr (1975) in Denmark and The

Figure 3.8 Transect and time–space diagram from Hallet's Inlet, Long Island, New York. (a) Stratigraphy of sand, silt and peat layers in sediment cores relative to topography in a transect across a tidal flat. (b) Summary of changes in vegetation based upon 32 pollen samples (indicated by solid squares). Plant assemblage symbols are: A = emergent fresh water aquatic; C = Cyperaceae; F = high-marsh fringe; S = salt marsh; T = shrub thicket. From Clark (1986).

Netherlands respectively. Iversen (1969) documented the mid-Holocene changeover from a mull soil, characterized by a high cation-exchange capacity, relatively high pH, and evidence of biological activity that included mixing and aeration by earthworms, to a mor soil, with low pH and little biological activity and with humus accumulation. Pollen assemblages from throughout the soil profile showed that a major change in forest composition occurred with the change in soil conditions, from a mesic deciduous forest dominated by lime (*Tilia cordata*), hazel (*Corylus*), and oak (*Quercus*), to a more depauperate forest of oak and birch (*Betula*). Fossil-pollen studies of a forest soil in Mantingerbos, The Netherlands (Stockmarr, 1975) produced similar results to those in Denmark. In the Dutch study, the transition from mull to mor soil was demonstrated to occur following the first human disturbance of the forest about 1000 years ago. The mull soil was developed on brook sand, and initially supported a forest of *Tilia*, *Quercus* and *Corylus*. After initial cutting of trees, beech (*Fagus sylvatica*) and *Betula* increased in importance within the forest. Tree-cutting activity intensified in medieval times, and resulted in conversion to an open forest of *Fagus*, *Ilex*, *Sorbus*, and *Carpinus*.

These studies emphasize the role of allogenic factors, such as progressive soil leaching and human interference, in effecting long-term changes in the direction of successional processes away from the typical mesic, late-successional vegetation that might be predicted on the basis of autogenic factors alone.

3.5 CYCLES OF SECONDARY SUCCESSION

Recurrent disturbances such as repeated fires continually reset the trajectory of plant succession (Loucks, 1970; White, 1979; Delcourt *et al.*, 1983). A series of such disturbance events, each followed by vegetation recovery, can serve as 'natural experiments' that can allow testing of whether the direction and sequence of plant succession is repeatable, or predictable over a series of cycles.

3.5.1 Repeated disturbances

In order to use lake sediments to recover paleoecological information pertinent to the question of the predictability of plant succession after episodes of repeated disturbance, several conditions must be met. First, the rate of sediment accumulation must be sufficiently rapid to record the disturbance events and their effects on vegetation in successional time. Second, the sedimentary sequence must preserve appropriate forms of evidence that can be sampled and interpreted ecologically. In the case of fire history, for example, particles of charcoal produced in each fire event are carried both by air currents and by sheetwash erosion from the upland soil surface into the lake basin (Patterson *et al.*, 1987; Clark, 1988a, b, 1990). Fossil pollen assemblages constitute an independent sample of former vegetation before, during, and after each fire event. A third requirement for quantitative time-series analysis to identify cycles of succession is the ability to take many samples equally spaced in time (Green, 1981; Birks and Gordon, 1985). Under ideal circumstances, lake sediments deposited as annual layers (varves) provide appropriate time resolution as well as insuring that no significant mixing of fossil assemblages has occurred between layers, mixing that could otherwise blur the signal of either disturbance episodes or vegetation response (Swain, 1973; Clark, 1988b). A final constraint that is critical for time-series analysis is that the series of data contain no overall trends in either statistical mean or variance (Green, 1981). For fossil pollen data, this means that only those portions of the paleoecological record may be used during which the overall composition of the vegetation is stable (as in trajectory A of Fig. 3.3). During times of transition from one major vegetation type to another, overall population increases or declines in principal taxa will invalidate the results of the time-series analysis (Green, 1982). However, time intervals occurring before and after transitional changes may be compared for identifying similarities and differences in both fire regimes and patterns of plant succession.

In a study of forest succession following repeated occurrence of fires in Nova Scotia, Green (1981) used sediments from Everitt Lake to analyze post-fire responses of individual tree taxa. Everitt Lake did not contain varved sediments, but it was a closed basin that was demonstrated by radiocarbon dating to have rapid and nearly uniform rates of sedimentation through the Holocene. During the time interval from 4450 to 2100 radiocarbon years ago, Green (1981) obtained charcoal and pollen samples at approximately 50-year intervals (taken at vertical intervals approximately 3 cm apart in the sediment core analyzed). Time-series analysis of pollen, charcoal, and inorganic influx rates for this interval revealed that the average fire cycle in Everitt Lake's drainage basin was 350 years in frequency (Fig. 3.10).

Postfire responses of different tree taxa were evaluated by statistical correlation of their response times for increase in populations following the fire events recorded by charcoal in the Everitt Lake sediments (Fig. 3.11). Most disturbance-favored taxa, such as pine (*Pinus*), spruce (*Picea*), larch (tamarack, *Larix laricina*), willow (*Salix*), and elm (*Ulmus*) (Fig. 3.11) showed significant increases in correlation at a +50-year time lag after fire. In general, late-successional trees such as beech (*Fagus grandifolia*), balsam fir (*Abies balsamea*), hop-hornbeam (*Carpinus/Ostrya*), and maple (*Acer*) had highest correlations much longer after fire events. Oak (*Quercus*) and ash (*Fraxinus*) had intermediate response times to fire (Fig. 3.11).

Green also examined patterns of successional replacement of tree taxa by examining the correspondence of postfire response between taxa. From this analysis, it was not possible to determine whether or not two or more taxa

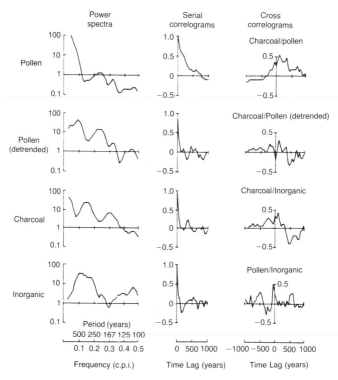

Figure 3.10   Power spectra, serial correlograms and cross correlograms for accumulation rates of pollen, charcoal and inorganic components of Holocene lake sediments from Everitt Lake, Nova Scotia, based on time series analysis. The power spectrum for charcoal shows a significant peak at 350 years that corresponds with a peak in accumulation rates of inorganic sediments associated with fire events in the lake's drainage basin; a second peak in the power spectrum for charcoal at 167 years reflects the frequency of fire independent erosion events. Comparison of the pollen/inorganic correlogram with the charcoal/pollen correlogram shows that the highest correlations occur at −50 yr time lag, implying that high erosion precedes high pollen influx, a pattern produced by fires in the lake's drainage basin. From Green (1981).

grew together in the same habitat within the watershed of the lake. However, Green was able to group assemblages of trees that would be expected to be common during various intervals of time following a major fire event (Fig. 3.12). In general, the results of this study were consistent with inferences made from known life-history characteristics of the taxa involved. Early-successional taxa such as poplar (*Populus*) and willow responded together in the first 50 years after fire; late-successional communities of hemlock (*Tsuga*), beech, and maple were the last to assemble, requiring 300 to 350 years after fire to establish mature communities (Fig. 3.12).

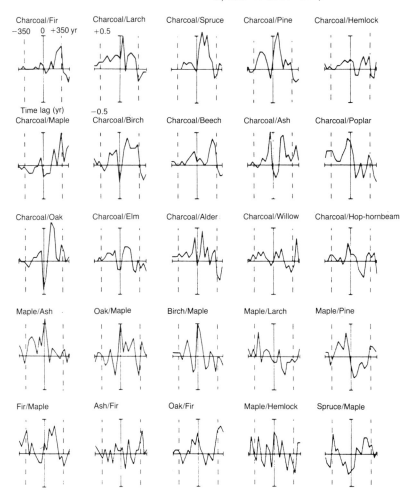

Figure 3.11 Cross correlograms of charcoal and pollen accumulation rates for the time period 4450 to 2100 yr B.P. at Everitt Lake, Nova Scotia. The vertical scales represent correlations (±0.5) and the horizontal scales represent time lags. Dashed lines mark one 350-year interval each side of zero years time lag. Changes in pollen accumulation rates after fire are indicated by the correlations at positive lags between zero lag of the fire event and the right dashed line. From Green (1981).

Comparison of results from analysis of a previous time interval, from 6600 to 4450 years ago, revealed that during the mid-Holocene, fire frequency was higher, with an average fire recurrence interval of 250 years. During that period, pollen influx rates of early-successional pine were higher and those of late-successional fir and beech were lower than during the later time interval. From these relationships, Green (1981) inferred that the course of plant

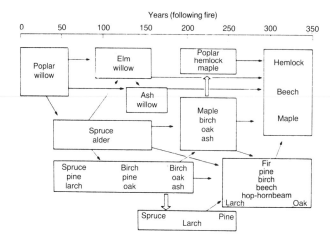

Figure 3.12 Successional trends in forests following fire episodes on the watershed around Everitt Lake, Nova Scotia, inferred from pollen data for the time interval 4450 to 2100 yr B.P. Taxa grouped within boxes are interpreted as plants typically growing together. Black arrows indicate directions of replacement of tree taxa in forest succession. Open arrows indicate disturbances. From Green (1981).

succession was predictable in the assemblages of trees that tended to occur together in postfire communities of specified age since disturbance, but that the recurrence interval of the natural fire regime was a major allogenic factor in determining the extent to which forest succession was allowed to proceed between recurrent disturbance events.

3.5.2 Cyclic autogenic regeneration in peatlands

Peter White (1979) contended that even in cases where autogenic processes control vegetation development, it is possible to identify some underlying environmental factor which is a stimulant to biotic responses that result in specific vegetation patterns. The question then becomes: is it possible to find any examples of cyclic successional change that are attributable to autogenic processes that can be demonstrated to occur in the absence of disturbances or other allogenic factors? One system for which critical examination of this question has been approached using paleoecological techniques in that of bog formation (Aaby, 1976; Barber, 1981; Payette, 1988; Foster *et al.*, 1988).

In peatlands, cyclic autogenic regeneration has been hypothesized to be induced by changes in ecological conditions resulting from a developmental sequence of different taxa and their growth phases (Payette, 1988). This form of succession was originally described for European raised bogs in which peat growth and hummock formation alternates with development of water-filled hollows (Barber, 1981; Payette, 1988). The concept of the autogenic 'regen-

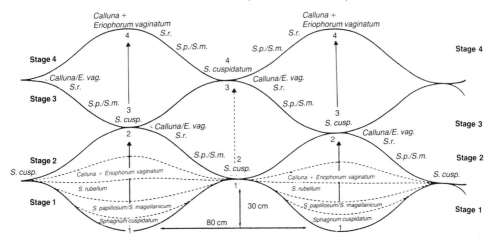

Figure 3.13 Idealized regeneration cycle of a peat bog in England reflecting alternation development of *Calluna* and *Eriophorum* hummocks and hollows (pools of water) with species of *Sphagnum*. From Barber (1981).

eration cycle' within peatland as developed by von Post and Sernander (1910), Osvald (1923), and Tansley (1939) is as follows (Barber, 1981; Fig. 3.13). In Stage 1 of peat build-up, a hummock of peat is flanked by two wet hollows; as the hollows fill with peat, the hummock, a community of heather (*Calluna*) and cottongrass (*Eriophorum vaginatum*), remains at its original level. Communities filling in the hollows are comprised predominantly of *Sphagnum*, which eventually grows above the level of the first hummock, creating two new hummocks while flooding the now lower area of the old hummock (Stage 2 in Fig. 3.13). With repetition of the sequence, the bog surface is maintained in a mosaic of hummocks and hollows, whereas a stratigraphic cross-section of peat appears as a series of lens-shaped structures. In the classical description (von Post and Sernander, 1910), lighter-brown, less humified peat represents the hollow phase and darker, more humified peat is produced in the hummock phase.

The cause of the development of hummock-hollow patterns through time and across space has been questioned by recent investigations of peat stratigraphy that demonstrate the importance of fluctuations in hydrological conditions (Foster *et al.*, 1988) and climate (Aaby, 1976; Barber, 1981) in controlling rates of peat growth and decay.

Bent Aaby (1976) demonstrated that raised bogs, which are dome-shaped accumulations of peat that obtain their entire supply of moisture directly from the atmosphere, exhibit cyclic patterns of growth that correspond with fluctuations in climate. He attributed changes in the degree of humification of peat (that is, decay of organic matter producing humic acids) in the Draved bog, Denmark, to reflect changes in precipitation and/or temperature that altered the rates of decomposition with a periodicity of about 260 years over the past 5500 years. Although the area of hummocks increased in dry periods and decreased in wet intervals, Aaby demonstrated with radiocarbon-dated peat sections that specific hummocks have remained in place for over 2500 years. On this basis, he questioned the explanation of autogenic succession as accounting for the observed patterns of peat growth and decay in raised bogs (Aaby, 1976).

Keith Barber (1981) carried out an extensive investigation of Bolton Fell Moss in England specifically to provide a paleoecological test of the theory of cyclic peat bog regeneration, through a detailed comparison of historic precipitation records and radiocarbon-dated peat sections extending back over the past 1860 years. Wet phases in bog surface conditions, as inferred from pollen and plant-macrofossil evidence, were found to coincide with several prominent wet climatic periods based on historical records (Fig. 3.14), occurring from A.D. 900 to 1100, A.D. 1320 to 1485, and A.D. 1745 to 1800 (the peak of the Little Ice Age). Based on this evidence, Barber (1981) considered that peat development at Bolton Fell Moss has been sensitive to climatic changes, at least over the last 2000 years; alternating formation of pools and broad dry surfaces on the peatland, as reflected by changing macrofossil assemblages and degree of humification of the peat, were attributed to changes in temperature and/or rainfall rather than to autogenic processes.

David Foster *et al.* (1988) and Foster and Wright (1990) compared the surface characteristics and development of a completely different type of peatland, the string-bog and fen system, which occurs today in boreal regions from the Hudson Bay lowlands and Labrador to Sweden, Finland, and the USSR. Climates of these regions are characterized by an excess of precipitation over evaporation. Foster *et al.* (1988) emphasized the importance of under-lying landform and hydrology on formation of the characteristic patterns of raised peatlands (strings) separated by elongated pools within these peat-

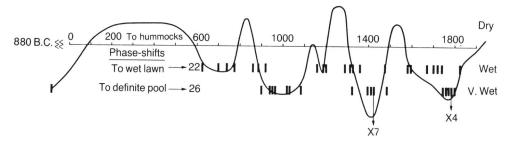

Figure 3.14   Changes in areal extent of peat hummocks and pools of water since A.D. 90 in Bolton Fell Moss peat bog, England. Increases in the relative height of the dry peaks reflect increases in both the area of peat hummocks and their degree of humification (decomposition of organic matter). From Barber (1981).

lands. Since the pools were found to be best developed on peat substrate and gradual slopes of the wetland terrain, Foster *et al.* (1988) suggested that the patterns in these fens are more directly controlled by biological processes in response to changes in hydrological conditions than by other physical mechanisms such as solifluction or by changes in atmospheric precipitation through time. In this situation, hollows initiate as unoriented small, low areas between hummocks. They deepen because of greater rates of peat accumulation around their margins and begin to coalesce as small hollows enlarge into broader pools, with pools tending to become aligned transverse to the direction of the slope. As the patterned peatland complex expands both laterally and vertically, the accumulating peat mass results in a general rise in water table through time. Ultimately, both surface and subsurface water flow results in dissection of the peat through piping and through stream erosion, leading to dissection of the peatland. Instability and change in peatland patterns in both Swedish and Labrador mires are initially related to allogenic environmental factors of landform and hydrology (Foster *et al.*, 1988). With time, the development of string-bog hummocks and pools is primarily the product of autogenic biological processes under hydrological control (Foster and Wright, 1990), but these hummocks and pools are not necessarily the result of cyclic responses of autogenic developmental processes.

Through a paleoecological investigation, Payette (1988) sought to determine whether autogenic succession could account for any of the pattern and process observed in the Clearwater Lake peatland in northern Quebec. To accomplish this, he first determined the long-term ecological changes represented in peat layers that were attributable to climate and local environmental variables. He then examined the record for evidence for short-term (100-year) repetitive alternations of the same plant species occurring over 1000-year time intervals of relative constancy in environmental conditions that would indicate biological replacement processes that were independent of external environmental controls (Payette, 1988). In the Clearwater Lake peatland, layers of *Sphagnum* peat were found to alternate with those of spruce (*Picea*) detritus over a period of the last several thousand years. Overall, through the past 5000 years of bog development, climate has been the primary control on floristic composition; however, within this time frame, alternating shifts in *Sphagnum* versus *Picea* may be attributable to internal peat-building processes in addition to the long-term response of the whole peatland system to climatic change (Payette, 1988).

Within the radiocarbon-dated peat sections studied by Payette (1988), the layers of *Sphagnum* and *Picea* peat were variable in morphological pattern, and cycles involving change from one community type to the other were continuous, short-term, and non-periodic over a 2300-year interval examined between 4010 and 1710 years ago. Replacement of *Sphagnum* by *Picea* is facilitated because the moss provides a medium for spruce layering; in turn, stunted spruce krummholz provides a favorable microclimate for increased snow accumulation for maintaining *Sphagnum* growth. Payette (1988) concluded that during the late Holocene, the continuously cold, moist subarctic climate of northern Quebec, combined with the lack of disturbance of the vegetation by fire, provided a long span of time in which the Clearwater Lake peatland was dominated by continuous, autogenic cycling of *Sphagnum* and spruce. Payette (1988) acknowledged that external controls including climate, landform, and hydrology vary geographically and lead to different long-term successional trends in different regions. He emphasized, however, that short-term vegetation cycles need to be assessed within the context of a long-term time scale in order to distinguish between allogenic and autogenic processes.

3.6   ANTHROPOGENIC INFLUENCES

Changes in plant communities invading old-fields following abandonment of agricultural land has provided a major source of ecological data on secondary succession, particularly as documented for the eastern United States (Odum, 1969; Marks, 1983). The trends in changing species composition, biomass, energy flow, and dynamics of resource allocation observed in old-field habitats, however, are historic patterns that have developed only within the last few hundred years following conversion of forests to cultivation. The accompanying fragmentation of the landscape has resulted in a patchwork whose configuration today is largely an artifact of modern land use (Burgess and Sharpe, 1981). Much of the flora today available for colonization of fallow fields is comprised of exotic taxa that have been introduced from Europe, Asia, or Latin America (Marks, 1983; Crosby, 1986). Ecosystem dynamics observed in such circumstances (Odum, 1969; Marks, 1983; Tilman, 1988) may be very different from those occurring in the absence of anthropogenic influences, because the taxa involved in the processes of old-field succession are a mixture of native and non-native species that are not highly co-evolved through time, but instead have only recently assembled into new, anthropogenically influenced communities.

3.6.1 Prehistoric origin of old-field succession

Peter Marks (1983) estimated that only two-thirds of the field plants in the northeastern United States (about 100 species) are native. From the autecology of these species, he concluded that most were not capable of persisting in the seed bank within mature forest and regenerating on the forest floor in temporary light gaps; rather, in presettlement times the majority of today's native old-field plants were adapted to persistent openings in the forest such as occur along streams and cliff edges within predominantly forested regions. Marks inferred that these species of herbs and shrubs were well adapted to their new roles in old-field succession because of their tolerance of full sun and their more effective seed dispersal characteristics than possessed by forest understory plants. However, Marks noted that many of the old-field species cannot be viewed as early-successional under all circumstances because they are not effective colonizers of ephemeral disturbance patches within forest communities and because they presumably evolved their life history traits in environments where their populations would have been persistent rather than ephemeral.

Marks (1983) dismissed the possibility of Holocene evolution of the species within prehistoric Indian old-fields, under the assumption that those habitats were also of relatively recent origin and would not account for the prior evolutionary history of the species. Evaluation of this additional factor is possible by combining evidence from the paleoethnobotanical record (remains of edible plants and wood used for fuel and habitations from archaeological sites) with that from local paleoecological sites within areas known to have been impacted by prehistoric human activities (Behre, 1986; McAndrews, 1988). From the archaeological and paleoecological record available across eastern North America (Delcourt, 1987), it can be inferred that prehistoric human activities affected the natural vegetation in several fundamental ways over a time span encompassing as much as the past 10 000 years (the entirety of the present interglacial interval). Prehistoric Native Americans (1) changed the dominance structure within forest communities (McAndrews, 1988); (2) extended or truncated the distributional ranges of both woody and herbaceous plant species (King, 1985); (3) provided disturbed open areas including agricultural old-fields into which native ruderals invaded and subsequently became weeds (Delcourt *et al.*, 1986a); and (4) changed the proportion of forested to non-forested land through time, progressively creating a culturally maintained landscape mosaic (Delcourt and Delcourt, 1988).

In eastern North America, much of the direct impact on the landscape mosaic caused by the activity of prehistoric Native Americans occurred along riverine corridors of the lower midwestern and southeastern United States, especially along major drainageways such as the Mississippi, Missouri, Ohio and Tennessee rivers and their tributaries (Delcourt, 1987). For example, in the Little Tennessee River Valley of East Tennessee (Delcourt *et al.*, 1986a), continuous Indian occupation is documented by the archaeological record for at least the last 10 000 years. The extent of area within the Holocene floodplain and adjacent Quaternary stream terraces of the Little Tennessee River that was disturbed by human settlement increased through time (Fig. 3.15), as human populations grew and as cultural changes took place that included introduction of new lithic and ceramic technologies in addition to cultivation of both indigenous and introduced crops (Delcourt, 1987; Delcourt and Delcourt, 1988). Ragweed (*Ambrosia*), considered by Marks (1983) to exemplify an indigenous old-field plant that would have been

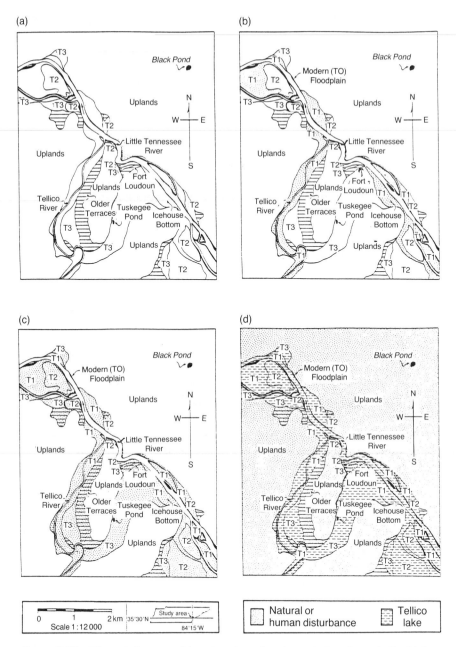

Figure 3.15 Changes in area of forest and nonforest vegetation through the Holocene on the watershed of the Little Tennessee River, East Tennessee: (a) PaleoIndian cultural period, 12 000 to 10 000 yr B.P.; (b) Archaic cultural period, 10 000 to 2800 yr B.P.; (c) Woodland and Mississippian cultural periods, 2800 yr B.P. to 500 yr B.P.; (d) Historic period, 500 yr B.P. to present, showing the current area inundated by Tellico Lake reservoir. T1 to T3 are late-Quaternary stream terraces. Nonforest vegetation was initially controlled in distribution by natural disturbances but increased in area through time because of prehistoric and historic human activities and land use. From Delcourt and Delcourt (1988).

confined to marginal habitats prior to 500 years ago, was represented as early as 1500 years ago by as much as 40% of the total pollen assemblage from Tuskegee Pond, a paleoecological site within the Little Tennessee River Valley (Delcourt *et al.*, 1986a). Tuskegee Pond is located on a broad stream terrace well beyond the limits of the active floodplain of the Little Tennessee River, a location that was densely forested in the early Holocene and progressively cleared and cultivated during the middle and late Holocene intervals (Fig. 3.15). At Tuskegee Pond, high prehistoric values of ragweed pollen occurred along with up to 2% maize (*Zea mays*) pollen (indicating local cultivation of corn) and representation of other cultigens and ruderals including goosefoot (*Chenopodium*), sumpweed (*Iva annua*) and purslane (*Portulaca oleracea*) (Delcourt *et al.*, 1986a).

Prior to 10 000 years ago ecological relationships differed from those that developed during the Holocene, for example including interactions of plants and now-extinct large herbivores and grazers (Janzen and Martin, 1982; Graham, 1986). Evolution of many of the plant species now characteristic of abandoned agricultural old-fields may have taken place much earlier during previous glacial or interglacial intervals in which the vegetation mosaic was very different from that of the Holocene, and particularly different from that immediately preceding the colonization of the North American continent by EuroAmericans some 500 years ago. The Holocene has been a time interval of progressive human influence on the vegetation mosaic, and many of today's native old-field plant species can be demonstrated to have occurred within anthropogenically derived open areas over at least the last 5000 years (Delcourt *et al.*, 1986a).

3.6.2 Historic trends in succession

Successional changes over the past 100 years can be traced effectively using paleoecological techniques, provided that a detailed and independent chronology is available. Within this time scale, varved sediments may be used to give precise dates. In the absence of varves, fossil-pollen evidence can be correlated with local land use history using known events that make distinctive changes in the pollen record. An example of such a pollen 'time line' is the well-documented decline in populations of the American chestnut (*Castanea dentata*), which occurred over much of eastern North America during several decades early in this century (Anderson, 1974). Another method of independently dating events during historic time is that of Pb-210 dating, using an isotope of lead that has a half-life of 22 years (Brugam, 1978a, b).

In a classic study, Dick Brugam (1978a, b) applied the techniques of radiocarbon dating and Pb-210 dating to historic sediments of Linsley Pond, Connecticut. He developed an absolute and independent chronology with which he could compare events within the pollen record to records of changing land use in the vicinity of the pond in order to reconstruct changes in plant succession since the time of Euro-American settlement.

The pollen percentage diagram from a 200-cm core of Linsley Pond sediments spans the time interval from about A.D. 1200 to present (Fig. 3.16). Presettlement forests of southern Connecticut were a mixture of oak

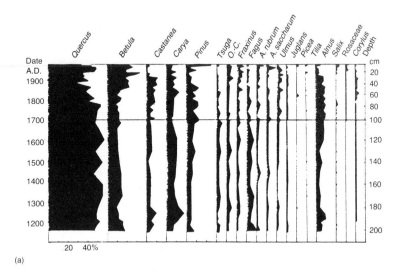

(a)

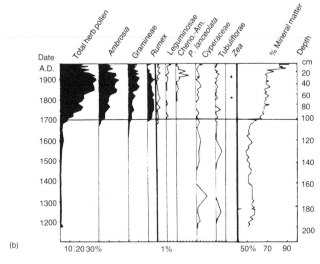

(b)

Figure 3.16   (a) Percentage diagram of tree pollen from a sediment core of Linsley Pond, Connecticut. *O.–C.* = *Ostrya/Carpinus* type. Horizontal line at A.D. 1700 represents the time of EuroAmerican settlement. (b) Percentage diagram of herb pollen and mineral sediments from the Linsley Pond sediment core. From Brugam (1978a).

(*Quercus*), birch (*Betula*), American chestnut (*Castanea dentata*) and hickory (*Carya*), with minor representation of pine (*Pinus*), hemlock (*Tsuga*) and mesic hardwoods. Closed forest surrounded Linsley Pond, as indicated by <5% total herb percentages. After A.D. 1700 total herb pollen increased dramatically, with primary contributors including ragweed (*Ambrosia*), grass (Gramineae), and dock (*Rumex*). Pollen of corn (*Zea mays*) was also present. The percentage of mineral matter in the sediments increased, reflecting deposition in the lake associated with increased rates of erosion generated by activities of homesteading and local cultivation. The presence of *Rumex* pollen was interpreted (Brugam, 1978a) as evidence for the maintenance of pastureland near the lake from the time of the arrival of the Linsley family.

Little change in percentages of pollen of tree taxa occurred at the time of first settlement near Linsley Pond, probably because the first homesteaders cleared only local areas for immediate needs (Brugam, 1978a). However, after A.D. 1850 an increase in birch was attributed to an increase in land abandonment as marginal farmland was allowed to return to forest. The chestnut blight (*Endothia parasitica*) arrived in southern Connecticut about A.D. 1910 and was registered by a decline in *Castanea* pollen in the Linsley Pond pollen sedimentary record. After the chestnut decline, an increase in birch pollen occurred over about 20 years, followed by a sharp decline. Subsequently, oak increased, inferred by Brugam (1978a) as a successional sequence from birch to oak on sites originally occupied by chestnut trees. Recent afforestation with pine is reflected in the large increase in pine pollen in the interval from the 1960s to 1970s.

The Linsley Pond study (Brugam, 1978a, b) demonstrates that together, Pb-210 analysis and radiocarbon dating can provide an accurate chronology for the past millennium that permits detailed comparison of presettlement vegetation with historical records of land use history and with postsettlement changes in plant succession.

3.7 CONCLUSIONS

1. In order to determine the repeatability or predictability of successional trends observed across the modern landscape it is necessary to examine plant succession within the context of a longer time frame that includes many cycles of succession. New insights can be gained through paleoecological analysis of repeated sequences of disturbance events followed by plant succession through a long time-series of data.

2. Paleoecological techniques can be used to isolate intervals of relative constancy in climate in areas with minimal natural disturbance regimes in order to test the question of whether plant succession is always dictated by allogenic factors, or whether under certain circumstances autogenic succession is responsible for observed patterns and processes. The majority of paleoeco-

logical evidence indicates that allogenic factors including changes in climate, hydrology, and disturbances such as fire constitute the predominant controls over plant succession. Evidence for autogenic succession can be demonstrated only in cases in which climate and hydrology are relatively constant and disturbances are minimal.

3. Both point-source disturbances such as wildfire and broad-scale alterations in environmental conditions, including regional and global climatic changes, influence plant succession over both the short term and the long term.

4. Patterns and processes of secondary succession on most modern landscapes are in large part an artifact of anthropogenic disturbance; caution should be exercised in extrapolating conclusions about the evolutionary history of life-history traits based upon the responses of species to circumstances that have undergone radical changes in only the last several hundred years to several thousand years of human influence.

4 Gradients, continua and ecotones

4.1 ISSUES

Patterning in biotic communities can be discerned on the landscape as patchiness in the distribution of species populations, leading for example to mosaics of vegetation of differing appearance in terms of life form, or physiognomy. Where distributions of species populations overlap over a relatively broad geographic area, mappable communities may be apparent as recurring species assemblages on sites with similar aspect, slope, soil moisture, or soil type. A long-standing controversy in ecology has been whether such communities as found today represent associations of plants and animals that evolved together through time or whether they are relatively recent assemblages whose distributions coincide because of environmental and historical factors. The Clementsian view of plant community associations that remain intact through time was dominant in American ecology during the first half of the 20th century (McIntosh, 1985), even though the alternative view, that species are distributed individualistically according to their tolerances, was also proposed early in the history of the field of ecology (Gleason, 1926).

This controversy was partially resolved by studies of gradient analysis published largely in the 1950s (Whittaker, 1956; Curtis, 1959). These studies indicated that, although disturbances and environmental discontinuities may bound the local distributions of species into mappable community patterns, nevertheless along continuous and smooth environmental gradients the species tend to array themselves with population maxima situated at different environmental optima and with no two distributions exactly alike (Whittaker, 1975). For many years, the accepted explanation for the existence of these ecoclines, based entirely upon data from modern and presumed static environmental gradients, has been that evolutionary processes act on species populations so as to select for different optimal portions of the gradient, thereby minimizing competition and niche overlap (Whittaker, 1975).

In order to consider the processes that are operative in both ecological and evolutionary time, it is necessary to ask whether the position, steepness, and length of environmental gradients and the overlapping species populations as arrayed today have remained constant in time and space, or whether these ecoclines have been variable in location and extent (Delcourt *et al.*, 1983; Graham and Lundelius, 1984). This question has a major bearing on the question of the discreteness of biotic communities through time as well as across space. Once disassembled because of major environmental changes such as climatic change, do communities of plants and animals tend to reassemble in similar combinations, or does community composition vary as the species respond individualistically to environmental changes? Are the resulting changes in community composition predictable or are they random? These questions have been approached recently by Quaternary ecologists, both in examining the fossil plant record (Cole, 1982, 1985; M. Davis, 1976, 1981a, b, 1983; Delcourt and Delcourt, 1987a) and the faunal record (Graham, 1976; Graham and Lundelius, 1984). In these studies, actual changes in the distributions of plant and animal taxa have been mapped during the last transition from glacial to interglacial conditions (about 18 000 to 10 000 yr B.P.), the time of greatest magnitude of change in climate and biota in the past 100 000 years (Ruddiman and Wright, 1987).

4.2   GRADIENT ANALYSIS AND THE CONTINUUM CONCEPT

Several approaches have been developed for gradient analysis using paleoecological data. Over very long environmental gradients, spanning the subcontinent of eastern North America, fossil samples of pollen or of vertebrate remains have been recovered from transects of sites that span part or all of the past 20 000 years. These paleoecological data have been used to examine changing continua both across space and through time (Graham, 1976; Delcourt and Delcourt, 1987a). On single watersheds, studies of vegetation changes have provided insight into changes in steepness of environmental gradients through time in response to climatic changes (Delcourt *et al.*, 1983). The concept of the vegetational continuum across space has been expanded to include the dimension of time (White, 1979); vegetational change at any location through time is a continuum of more or less gradual changes in species composition and abundance (Jacobson and Grimm, 1986).

4.2.1 Patterns of species diversity of small mammals

Russell Graham (1976) compared the species richness of the vertebrate fauna during the late Pleistocene with that of the Holocene along a major northeast-to-southwest transect from just south of the full-glacial ice margin in Pennsylvania to the Great Plains of central Texas. In this study, Graham was interested in testing the hypothesis that a Pleistocene climate more equable than today could account for observed gradients in distribution of faunal communities as recovered in the fossil record. Relatively steep environmental gradients during the present interglacial interval are reflected in patterns of species occurrence and species richness. Northern range limits of warm-temperate taxa of small mammals found today in the eastern deciduous forest region are largely determined by winter temperature extremes, and southern range limits of boreal taxa are truncated by combinations of summer temperature extremes. Small mammals living today in the Great Plains are limited in their eastern ranges by available moisture (Graham, 1976).

Vertebrate communities were very different in composition during the Pleistocene, both because a large number of forms in all size classes became extinct at the end of the Pleistocene (Graham and Lundelius, 1984), and because the extant species occurred together in assemblages no longer found together today. These Pleistocene faunal assemblages were considered by Graham (1976) to represent an intermingling of populations along a much shortened environmental gradient compared with that of today. For both voles and shrews (Fig. 4.1), Graham found that species richness reflects environmental conditions, with greatest species richness both today and in Pleistocene times occurring at localities in northeastern North America, and with least species richness in the southwest, reflecting the southwestward trend toward both increasing summer temperatures and decreasing effective moisture. Total species richness was generally higher during the Pleistocene at all localities, inferred by Graham (1976) to reflect a more equable late Pleistocene climate with reduced gradients in temperature and moisture. From this study, Graham (1976) concluded that during the present postglacial interval, increasing seasonality of climate accentuated climatic gradients and thereby narrowed ecotones between vegetation types, influencing both the present distributional patterns and species richness of the mammalian fauna.

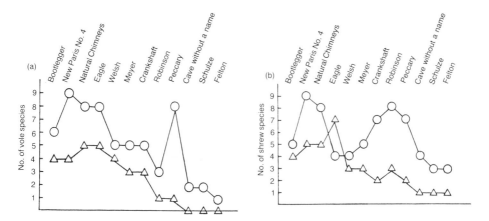

Figure 4.1 (a) Species richness of voles in faunas of late Wisconsinan age (circles) and modern age (triangles). (b) Species richness of shrews in faunas of late Wisconsinan age (circles) and modern age (triangles). The fossil-mammal sites are arrayed geographically from northeast to southwest across eastern North America and they are presented from left to right on the diagrams. From Graham (1976).

4.2.2 Changing gradients on a single watershed

Paleoecological data can be used to infer changes in environmental gradients on individual watersheds that in turn affect both the position and steepness of ecotones, that is, transition zones between different community types. Because of differential production and dispersal of pollen grains by taxa of gymnosperms and angiosperms, it is not possible to reconstruct precisely the population distributions of individual species across a watershed on the basis of fossil pollen samples from the center of one pond. However, a number of complementary forms of information provide indices that can be used to infer changes in hydrology and soil moisture gradients that, in turn, affect overall community composition on the watershed. These data include changes in type and rate of accumulation of both organic and inorganic sediments in the pond, as well as the composition of aquatic macrophytes and of upland plants inferred from changes in both pollen and plant-macrofossil assemblages through the sediment sequence. From the relationship of the pond surface area to the area of its watershed the general pollen source area can be inferred in order to estimate the area over which plant communities have changed in location and composition through time (Prentice, 1988).

The late-Quaternary paleoecological record from Cahaba Pond, Alabama, exemplifies the use of this approach in reconstructing changes in vegetation,

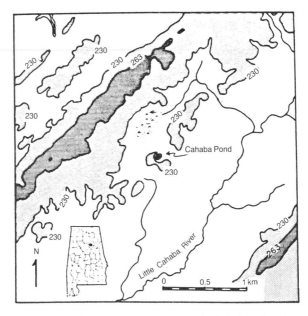

Figure 4.2 Location and topographic map of Cahaba Pond, north-central Alabama. Contour elevations expressed in meters. From Delcourt *et al.* (1983).

water levels in the pond, and dynamics of ecotones between plant communities through the Holocene interglacial interval (Delcourt *et al.*, 1983). The Cahaba Pond watershed (Fig. 4.2) represents a microcosm within the Ridge and Valley physiographic province of north-central Alabama. The pond is 0.2 hectares in surface area, and it originated 12 000 years ago as a sinkhole collapse in underlying Ordovician limestone. The upper slopes of the Cahaba Pond watershed are underlain by Pennsylvanian-age sandstone that is relatively resistant to weathering. The pond is springfed from a shallow aquifer. Water level fluctuations are sensitive to changes in total rainfall and its distribution throughout the year. Vegetation today occupying the Cahaba Pond watershed ranges from lowland red maple (*Acer rubrum*) forest, through midslope mixed deciduous forest including hickory (*Carya*), ash (*Fraxinus*), sweetgum (*Liquidambar styraciflua*), flowering dogwood (*Cornus florida*), and numerous species of oak (*Quercus*), to driest upper slopes on sandstone substrate that are vegetated by a forest stand of loblolly pine (*Pinus taeda*) and Virginia pine (*Pinus virginiana*).

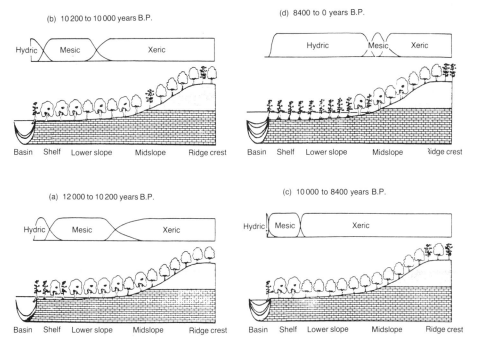

Figure 4.3 Idealized topographic profiles for the Cahaba Pond watershed, illustrating Holocene changes in pond area, water depth, patterns of sediment accumulation, and the distributions of hydric, mesic, and xeric vegetation. Positions and relative widths of ecotones between community types are speculative. From Delcourt *et al.* (1983).

Gradients in both elevation and underlying substrate are pronounced on the Cahaba Pond watershed. Inferences about changes in these gradients (Fig. 4.3) were based on paleoecological data that included analysis of changes in both aquatic macrophytes (sensitive indicators of pond-water levels) and in pollen and plant macrofossils representing vegetation growing from lower slope to ridgetop. From 12 000 years ago to 10 200 years ago (Fig. 4.3(a)), a mesic, primarily deciduous forest dominated by beech (*Fagus grandifolia*) and oaks surrounded Cahaba Pond, with at least 75% of the tree pollen produced by hydric and mesic taxa. During this time interval, the watershed was dominated by a stand of beech trees. Trees tolerant of a long period of high water levels during the year, such as Atlantic white cedar (*Chamaecyparis thyoides*) and bald cypress (*Taxodium distichum*), occupied the pond margins. Species of oaks and hickories, which are generally more tolerant of drought stress, grew primarily on the xeric ridgetop; broad ecotones probably existed between most communities on the watershed of Cahaba Pond during this time interval of equable environment.

Between 10 200 and 10 000 years ago (Fig. 4.3(b)), rapid vegetational changes occurred as the local population of beech died back and was replaced by a succession of tree taxa including cucumber magnolia (*Magnolia acuminata*), white pine (*Pinus strobus*), hemlock (*Tsuga canadensis*), and chestnut (*Castanea dentata*). Pollen accumulation rates for most mesic taxa decreased markedly, indicating the demise of their populations and the onset of increasingly dry climatic conditions. Oak and pine increased, presumably on the hillcrest as well as generally throughout the Ridge and Valley region. Cedar and cypress were eliminated from the local watershed, indicating that water levels in the pond had dropped, exposing a broad limestone shelf. With climatic drying in the early Holocene, populations of submersed aquatic plants such as mermaid weed (*Najas*) disappeared and were replaced by shallow-rooted, floating-leaved and emergent aquatic plants.

From 10 000 to 8400 years ago (Fig. 4.3 (c)), oak and hickory became predominant in the fossil pollen record, and the representation of mesic trees diminished. Water levels in the pond were lower than previously, with peat-forming, emergent sedges and rushes (family Cyperaceae) and damp-ground plant species replacing aquatic plants in the plant macrofossil record. During this time interval, climatic warming and drying may have resulted in both a steepened gradient in soil moisture on upland slopes and sharpened ecotones between forest communities because of increased evaporation from the ground surface and increased transpiration from the plants.

After 8400 years ago (Fig. 4.3(d)), black gum (*Nyssa sylvatica*) became established locally, along with wetland shrubs such as buttonbush (*Cephalanthus occidentalis*) that tend to trap sediments in the vicinity of their root systems. The pond had filled in with sediment, and water levels once again rose and inundated the limestone shelf and lower slope of the watershed. With climatic amelioration, the modern vegetation patterns developed. Southern pines established locally on ridge crests, and ecotones between communities broadened once again (Delcourt *et al.*, 1983).

The example from Cahaba Pond is instructive in illustrating the sensitivity of local watersheds to environmental changes. Rather than providing a stable substrate and fixed environmental gradients as a setting for long-term ecological and evolutionary processes to operate (Swanson *et al.*, 1988), both physical gradients and biotic communities on this watershed underwent considerable change in response to even subtle climatic changes in the supply of precipitation relative to its loss through evaporation and transpiration during the span of an interglacial interval.

4.2.3 Rates of vegetation change – continua through time

In a study of the Holocene history of vegetation at Billy's Lake, located in central Minnesota, George Jacobson and Eric Grimm (1986) demonstrated that vegetational composition through an interglacial time interval at a site can exhibit continual change with relatively few intervals of constancy or of minimal change. The vegetation continuum documented through the past 10 000 years at Billy's Lake was similar, though not exactly the same in sequencing, to vegetational continua that today can be traced across a spatial transect from forest to prairie in Minnesota (McAndrews, 1966).

At Billy's Lake, located near the modern transition from forest to prairie in the western Great Lakes region, an early Holocene forest of jack and red pines (*Pinus banksiana* and *Pinus resinosa*) was replaced by prairie after 8000 years ago. Prairie was subsequently displaced by forest largely composed of white pine (*Pinus strobus*) after 1200 years ago. Changes in composition of plant communities throughout the 10 000-year history, however, were nearly continuous, with only one interval of relative stasis apparent from the fossil-pollen record, occurring between 7000 and 6000 years ago. The continuous nature of vegetational change was demonstrated by analysis of the pollen percentage data using Detrended Correspondence Analysis (Gauch, 1982), a technique of ecological ordination that determined the degree of dissimilarity between fossil pollen assemblages averaged in consecutive 100-year intervals and arrayed in a continuous series through time (Fig. 4.4). The greatest rate of change in vegetation represented by the fossil pollen assemblages occurred during times of transition from forest to prairie or from prairie to forest, with minimal change apparent during the middle-Holocene time when the area was occupied by prairie. The single 100-year time span of greatest vegetational change occurred in the last 100 years of EuroAmerican settlement and conversion of natural vegetation to cropland (Jacobson and Grimm, 1986).

From the long-term record, it is clear that vegetational continua arrayed

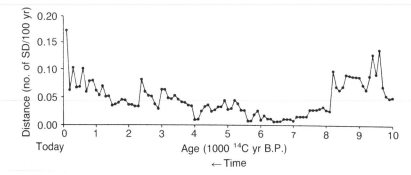

Figure 4.4 Rates of vegetational change in central Minnesota through the past 10 000 years based on ordination of fossil pollen spectra at Billy's Lake. On the y-axis, relatively large values of ordination distance indicate the times of relatively greater magnitude of change in pollen spectra per 100-year interval, and, therefore, the times of substantial vegetational change in the uplands surrounding Billy's Lake. From Jacobson and Grimm (1986).

across space must also change through time. Over an interglacial time span, the rate of change in vegetational continua is determined by changes in climate mediated through its effect on natural disturbance regimes such as fire (Grimm, 1983, 1984; Jacobson and Grimm, 1986). In an extension of the Billy's Lake analysis, the ecological ordination technique of Detrended Correspondence Analysis (Gauch, 1982) was used to examine rates of vegetational change in a series of sites located throughout eastern North America and extending back in time as much as 20 000 years (Jacobson *et al.*, 1987). In general, the times of greatest rates of vegetational change were found to occur primarily at the transition from Pleistocene to Holocene climatic conditions, between generally 12 000 and 10 000 years ago, as well as during the late-Holocene and historic time of the last 1000 to 100 years (Jacobson *et al.*, 1987). These studies indicate that inferences about the stability in distribution of vegetational gradients on the modern landscape or even in presettlement times must be viewed in the context of a longer-term trajectory of change. Historic plant communities are responding dramatically to the impact of historic humans, which has caused an acceleration in the rate of change in this temporal vegetational continuum.

4.3 ECOTONES AND ECOCLINES

Ecological transition zones, or ecotones, can be variously defined: floristically, by the presence or absence of species; by changes in structure or physiognomy of vegetation; or by changes in importance or dominance of taxa across a transition zone. The characteristics of ecotones between adjacent communities are in part determined by the scale of observation (Delcourt and Delcourt,

1991). Over an interval as short as a decade, landscape boundaries between adjacent communities on a hillslope may appear stable. In reality, however, the degree of stability of such an ecotone is dependent upon the magnitude and rapidity of disturbance events as well as the response time of the biota to environmental changes. Within natural forested landscapes, displacement of ecotone positions occurs only after a sustained and systematic change in mean annual temperature of about 200 years (M. Davis and Botkin, 1985). Over longer time scales of up to several thousand years, ecotones between large vegetation units, such as the prairie/forest border, may be displaced (Bernabo and Webb, 1977; Webb *et al.*, 1987).

4.3.1 Null models of ecotone dynamics

Generalities about the dynamics of ecotones cannot be made without considering possibilities at all scales in space and time (Delcourt and Delcourt, 1991). Hypothetical case studies, or null models, involving species distributional limits and processes of invasion and extinction, can serve as examples for which the rates and extents of change in ecotone position and strength are known and against which empirical data can be compared (Fig. 4.5). Numerical ordination techniques such as Detrended Correspondence Analysis (Gauch, 1982) can be used to provide a quantitative measure of changes in ecotone strength expressed as the rate of turnover in species populations across the ecotone. The rate of species turnover can be measured either in

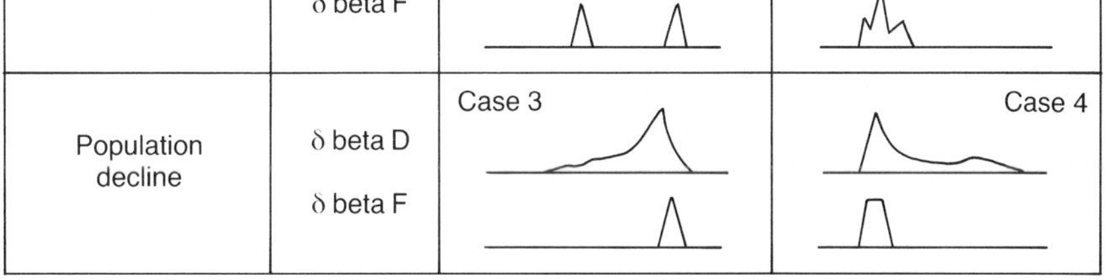

Figure 4.5 Idealized curves for rates of change in vegetational composition across ecotones with different dynamics. Curves for δ beta D are based on ordination of percentage (relative dominance) data across an ecotone using Detrended Correspondence Analysis (DCA); curves for δ beta F are based on ordination of presence–absence (floristic) data across the same ecotone using DCA. From Delcourt and Delcourt (1991).

terms of species composition (delta beta floristics, δ beta F in Fig. 4.5) or percentage dominance (delta beta dominance, δ beta D in Fig. 4.5) along a gradient through space or time. When total values of beta diversity are approximately 4.0 SD, they reflect a complete replacement of species assemblages along the ecotone (Gauch, 1982). Ordinations of data based on the presence or absence of species along a gradient (for example, floral or faunal species lists) identify the locations and breadth of significant transition zones at the mutual distributional limits of species. The locations of ecotones determined in this way differ from those based upon dominance data. The latter identify prominent ecotones only where large changes occur in the abundances of taxa. Use of two or more measures in combination (Fig. 4.5) gives the most information for delineating the overall breadth and strength of the ecotone (Delcourt and Delcourt, 1991).

Ecotone characteristics at one time can be compared with those in the same geographical location at another time, making it possible to quantify the dynamics of ecotone displacement. Given a long enough time-series of data from an appropriate observatory (Dyer *et al.*, 1988), such as either a Long-Term Ecological Research (LTER) site or a paleoecological site, the investigator can compare changes through time at one location with those occurring at a sequence of time frames across a given spatial transect. Simultaneous analysis of data sets across space and time reveals the most about ecotone dynamics.

An example of the null model approach for studying long-term ecotone dynamics is a case in which a major climatic change results in the decline and local extinction of a formerly dominant species, followed by its replacement in the community (Case 3 in Fig. 4.5; Fig. 4.6). In such a case we might think of the replacement phenomenon as a series of expansions of species populations on a watershed. Let us assume that all four species (A, B, C, and D in Fig. 4.6) are present at the outset, and that changes in their dominance values through time are analogous to the passage of an ecotone between boreal and cool-temperate vegetation through the site. In this hypothetical example and using percent dominance data, species A experiences a major population decline that results in temporary replacement by B prior to expansion of C and D along the gradient. The most important time of transition occurs during the changeover of species dominance during the declines of A and B (Fig. 4.6).

The usefulness of the analogy between vegetational change through time and ecotone displacements across space can be tested using the paleoecological record (Fig. 4.7). The case study from Anderson Pond, Tennessee (Delcourt, 1979), provides a situation in which, 20 000 years ago, the site was positioned near the full-glacial position of the transition zone between boreal and cool-temperate forests. Between 16 500 and 12 500 years ago, boreal forest dominated by jack pine (*Pinus banksiana*) and spruce (*Picea*) was replaced progressively by temperate deciduous forest during the prolonged interval of

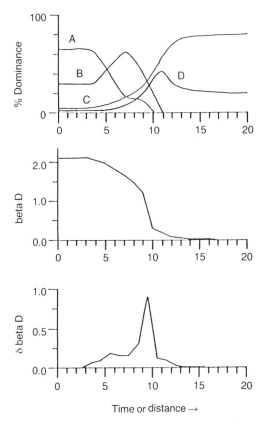

Figure 4.6 Hypothetical null model of species decline and replacement across an ecotone (Case 3 in Fig. 4.5). The ecotone may occur either along a transect of sites distributed across space or it may represent ecological changes through time at one site. From Delcourt and Delcourt (1991).

climatic amelioration in the late-glacial interval. Over the last 20 000 years there have been several times of major vegetational change corresponding with the passage of fundamental ecotone boundaries through the region. The strongest ecotone developed between 16 000 and 14 000 years ago (Delcourt and Delcourt, 1991). During that time interval jack pine populations were in decline because of climatic warming. Populations of both spruce and oak increased, in part because of the climatic change, and also in response to the loss of the formerly dominant jack pine. Although the sequence of replacement was much more complex in the paleoecological record from Anderson Pond than in the hypothetical case study, the pattern of replacement resembled in general outline the null model of population decline and replacement (Fig. 4.6).

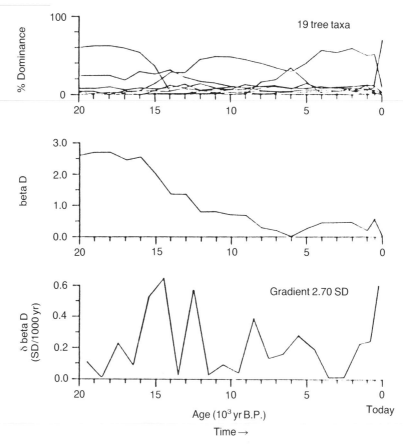

Figure 4.7 Analysis of ecotone dynamics from 20 000 yr B.P. to the present at Anderson Pond, Tennessee, based upon changes in percent dominance of tree taxa reconstructed from the pollen record. From Delcourt and Delcourt (1991).

4.3.2 Beta-diversity gradients in space and time

Ecotone analysis of sites along a transect at a number of different times in the progression of change from glacial to interglacial conditions can provide insight into the broad-scale dynamics of ecotones. In a study of long-term forest dynamics across eastern North America, we compiled mapped recon- structions of changes in distribution and dominance for 19 calibrated tree taxa (Delcourt and Delcourt, 1987a). We plotted ecoclines for all species, based on their percent dominance in reconstructed forests, along a transect that extended from the Gulf of Mexico to Hudson Bay at 85°W longitude. Ecocline data were ordinated at 4000-year intervals from 20 000 years ago to the present (Fig. 4.8). As the massive, continental Laurentide Ice Sheet,

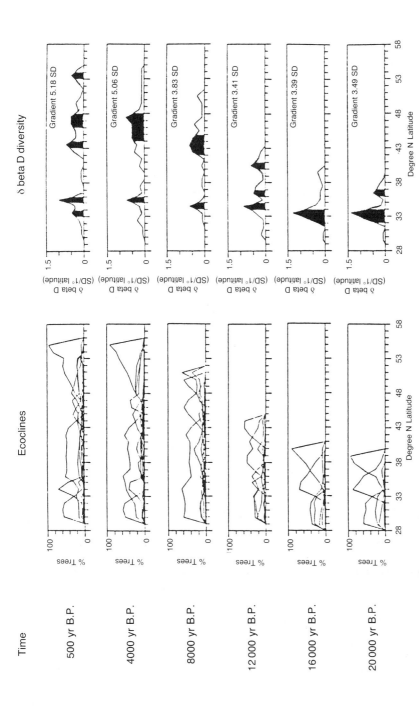

Figure 4.8 Late Quaternary ecoclines (including curves of percent dominance reconstructed from pollen records for 19 tree taxa) and ecotones (black bars on δ beta D graphs) between forest communities along a transect at 85° W longitude across eastern North America from 28° N latitude to 58° N latitude. From Delcourt and Delcourt (1987a,b); reprinted with permission from Pergamon Press.

which at its maximum extent during the late Wisconsinan interval of the late Pleistocene (20 000 years ago) extended south nearly to the confluence of the Ohio and Mississippi rivers, thinned and retreated northward during the late-glacial interval, greater land area became available for vegetation, and the length of the physical gradient increased accordingly. Through the time series, the total length of the beta diversity gradient also increased, probably as a result of the change from equable climatic conditions and relatively low turnover in species composition reflected in the few ecotones along the transect during full-glacial and late-glacial times to more seasonally contrasting Holocene climates, with higher turnover in species composition and increased species packing in a greater number of distinct plant communities.

The ecocline analysis of the 20 000-year time-series of paleoecological data across eastern North America leads to several key conclusions regarding the nature of ecotones and their dynamics (Delcourt and Delcourt, 1991). An ecotone may appear fixed in location through time (for example, the persistent transition zone between about 32°N and 34°N), and yet it may change in other attributes such as its strength as community composition changes across it. Given a strong enough environmental change, the number, position, and strength of ecotones between major vegetation types can all change. Different ecotones may have different sensitivities to the same environmental change, as a function of nearness to migration fronts for species populations. Ecotone dynamics may be different along different boundaries of the same community. If an environmental change causes a restructuring or disassembly of the communities that were responsible for definition of the ecotone, that ecotone may be ephemeral (Delcourt and Delcourt, 1991).

4.4 COMMUNITY DISASSEMBLY AND REASSEMBLY

The question of whether or not species respond individualistically to climatic or other environmental changes cannot be answered directly from examination of modern spatial distributions across environmental gradients. Studies that demonstrate differences among species in location of population centers and range limits, either along a topographic gradient (Whittaker, 1956, 1975) or across a broad region (Delcourt and Delcourt, 1981, 1984), indicate the existence of individualistic differences in tolerances to environmental thresholds, as well as the influence of biological competition in niche separation. However, changes in location of populations must be determined through time in order to draw conclusions about the degree to which communities remain intact or the extent to which they disassemble because of differential responses to environmental change. Further, a series of replicate experiments is required to approach the following question. Once disassembled, do communities tend to reassemble in the same way given a return to pre-existing environments, or are changes in community composition unpredictable and chaotic, dependent upon initial starting conditions as well as subsequent complex species interactions?

The Quaternary ecological record provides insight into the nature of community disassembly and reassembly. Climatic change has been a powerful environmental forcing function for biotic change over glacial/interglacial cycles, each cycle typically lasting for 100 000 years (Imbrie and Imbrie, 1979). Both biological and isotopic data from the marine sediment record indicate that as many as 20 glacial/interglacial cycles have occurred during the past two million years of the Quaternary Period (Ruddiman and Wright, 1987). Theoretically, fossil evidence from terrestrial contexts could provide replicate sequences over a series of glacial–interglacial cycles with which to test the null hypothesis that once disassembled, ecosystems tend to reassemble with similar structure, composition, and abundances of constituent species (Iversen, 1958; Kapp, 1977; Birks, 1986).

One example of late Pleistocene disassembly and Holocene reassembly of communities is illustrated by the studies of Ken Cole (1982, 1985) of rates of vegetational change based on packrat middens from northern Arizona. Preserved in rock shelters and caves in cliffs along the eastern wall of the Grand Canyon, packrat middens represent rodent collections of plant-fossil debris that are radiocarbon-dated from various times in the last glacial–interglacial cycle. Cole identified the species of the perennial plants represented as plant fossils within 52 midden samples. He determined temporal patterns of species richness and characterized overall flux of plant species in terms of arrival and departure (local extinction). Using an area-based sample of species available as food to foraging packrats, Cole identified the species of plants today occurring within a 30-meter radius of each packrat midden site. Cole then calculated the extent of taxonomic similarity between each fossil plant assemblage and the modern flora of each midden site. Values ranged from 0% similarity (no species in common) to 100% similarity (Fig. 4.9(a)). The full-glacial plant assemblages had the lowest values of percent similarity, averaging between 20% and 25% (Fig. 4.9(a)). The oldest plant-fossil assemblages are the most dissimilar to the modern flora growing at the same site. Plant-fossil assemblages dating from the middle and late Holocene, however, are quite similar (75% to 80%) to the modern flora. The greatest rate of change in similarity values occurred between 10 000 and 6000 years ago (Fig. 4.9(b)). Cole (1985) concluded that this early-Holocene convergence of the floral components reflected the assembly of modern plant communities by about 6000 years ago. A unidirectional change in vegetation is presumed because the data only reflect floristic changes relative to modern communities (Figs 4.9(a) and (b)). In order to determine the nature of ephemeral plant communities of the late-glacial and early-Holocene intervals, Cole (1985) calculated overall species flux (Fig. 4.9(c)), incorporating data for additional plant species that may have immigrated to, then emigrated from a site. He found that a peak in species flux occurred between 12 000 and 10 000

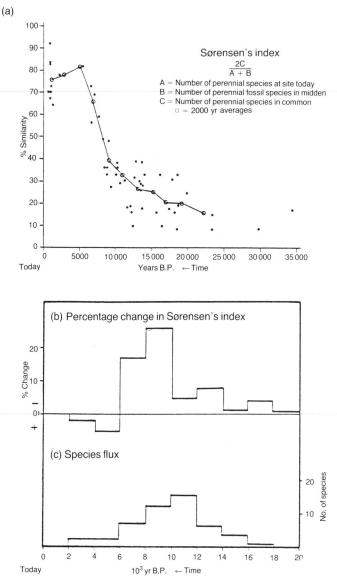

Figure 4.9 (a) Comparison of the similarity, using Sørensen's index, between the composition of plant macrofossil assemblages from late Quaternary packrat midden deposits with the modern composition of vegetation within 30 meters of each fossil site in the eastern Grand Canyon of Arizona. Open circles are values of Sørensen's index averaged into 2000-year intervals, with the solid line showing the trend between those averaged values. (b) Percentage change in Sørensen's index between 2000-year intervals, where the line equals the slope of the line in Figure 4.9(a). (c) Number of taxa making first or last appearance in the fossil record from the eastern Grand Canyon in each 2000-year interval. From Cole (1985).

years ago, substantially earlier than the primary interval of community assembly (10 000 to 6000 years ago). In order to determine the underlying cause of the late-glacial peak in species flux, Cole (1985) used the site elevation of each of the 52 packrat middens in order to separate the plant-fossil data into four altitudinal zones (midden sites located from 950 to 1220 meters, 1440 to 1470 meters, 1600 to 1900 meters, and 2000 to 2200 meters elevation). For each altitudinal zone, plant-fossil data for 'arrival' (invasion) and 'departure' (local extinction or emigration) were summarized in consecutive 2000-year intervals (Fig. 4.10(a)). Late Pleistocene plant communities lost a large number of perennial woody plant species abruptly between 12 000 and 10 000 years ago. New arrivals began to invade as early as 16 000 to 14 000 years ago, gradually increasing and peaking in terms of species flux of arrivals between 10 000 and 8000 years ago. The plant-fossil data for species number from the eastern Grand Canyon were plotted against radiocarbon age of sample for each midden (Fig. 4.10(b)) and as a smoothed curve plotted with a running average based on five consecutive samples (Fig. 4.10(c)). Both Pleistocene and Holocene plant communities maintained relatively high levels of taxonomic diversity. However, based on packrat midden samples from the eastern and western Grand Canyon (Fig 4.10(c) and (d)) and the Sheep Range in southern Nevada (Fig. 4.10(e)), Cole (1985) noted that a time interval of greatly reduced species richness occurred between 12 000 and 9000 years ago, during the transition from Pleistocene to Holocene conditions. Thus, plant species characteristic of Pleistocene communities were prone to local extinction prior to the arrival of Holocene species.

Cole (1985) interpreted the temporal patterns in species richness and flux within the context of Smith's (1965) model of 'biological inertia' of vegetational change. In the southwestern United States, community disassembly was triggered by major climatic and environmental changes that marked the end of Pleistocene. Cole (1985) postulated that the accelerated loss of species between 12 000 and 9000 years ago was the result of the postglacial climate shift which exceeded the climatic tolerance limits of many species and eliminated them from marginal habitats, producing a reduction in species richness in stressed sites such as along exposed cliff faces of the Grand Canyon. Species that were abundant during the late Pleistocene constricted their distributional ranges toward more favorable sites in their population centers. Cole (1985) asserted that these late-glacial communities persisted after the initial climatic change for another 1000 to 3000 years before new immigrants established local populations and outcompeted the original resident species. These persistent populations represented a form of inertia in the ecosystem, a kind of biological resistance to immediate climatically driven changes in plant communities. During the Holocene, modern plant communities assembled, continuing to accumulate new immigrating taxa until the floristic composition stabilized 6000 years ago. Changes in species

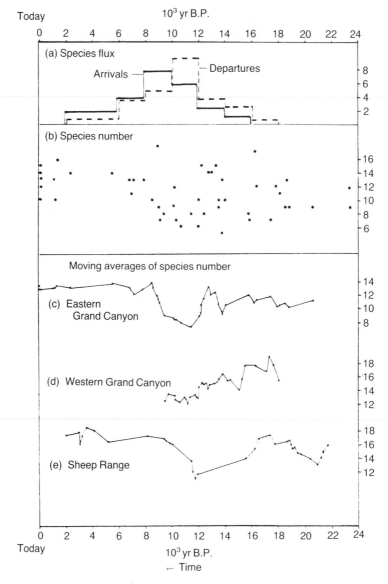

Figure 4.10   (a) Number of plant taxa making their first appearance (invasions recorded by the solid line) and last appearance (local extinctions recorded by the dashed line) in the eastern Grand Canyon during the late Quaternary, based upon plant macrofossils from packrat midden deposits. (b) Number of perennial plant species recorded in packrat middens from the eastern Grand Canyon. (c) Five-sample moving average of species number based upon plant–macrofossil data shown in Figure 4.10b. (d) Five-sample moving average of species number based upon plant–macrofossil data from the western Grand Canyon. (e) Five-sample moving average of species number based upon plant–macrofossil data from the Sheep Range of southern Nevada. From Cole (1985).

richness (Fig. 4.10(c)) illustrate the temporal pattern of community dis-assembly and reassembly. However, the influence of climatic change upon the lag in immigration of plant species during the Holocene remains a highly controversial topic that is the subject of ongoing research (Markgraf, 1986; Cole, 1986).

4.4.1 Community instability and coevolutionary disequilibrium

Among Quaternary ecologists, Bill Watts (1973) was the first to point out that traditional pollen zone subdivisions on diagrams of fossil pollen data em-phasize the times of relative stability in vegetation composition. These times of stability are separated by times of relatively rapid change, or instability, that in themselves have inherent interest to the plant ecologist but which require development of special means of analysis for proper ecological interpretation. Watts (1973) encouraged the examination of paleoecological data from the perspective of an understanding of the ecology of species populations for those taxa that are resolvable in the fossil record.

Margaret Davis (1986) has asserted that during times of major climatic change biotic communities will be in disequilibrium because of inherent time lags in response of different species that result from differences in behavior, life span, life-history characteristics, intrinsic rates of population increase and dispersal capabilities. Plant communities have inherently longer time lags for response than do animal communities, leading to changes in community structure that are apparent from analysis of the fossil record. Margaret Davis (1986) concluded that disruption of communities during times of major environmental changes prevents migration of communities as intact units.

Empirical evidence of community instability in the face of major climatic change is derived from three paleoecological sources: (1) replicate sequences of interglacial forest succession (West, 1961, 1977); (2) analysis of patterns of assembly of Holocene plant communities (Ritchie, 1985; Cole, 1985, 1986); and (3) evaluation of postglacial assembly of small mammal communities following extinction of a large number of vertebrate species at the end of the Pleistocene (Graham and Lundelius, 1984; Lomolino *et al.*, 1989).

From a series of organic-rich interglacial deposits in eastern England, West (1961) analyzed the pollen record to document four sequences of immigration of major plant taxa into the British Isles (Fig. 4.11; Wright, 1977). During each of the last four glacial periods, all trees were extirpated from Great Britain by extremely severe climatic conditions and development of a local ice cap (West, 1977). Tundra extended as far south as central France; primary refuge areas for temperate forest species were confined to the Mediterranean region south of the Pyrenees and the Alps (Huntley and Birks, 1983). In the paleoecological record from England (Fig. 4.11), the current interglacial interval, the Holocene, was characterized by initial dominance of boreal forest of birch and pine (West, 1977; Birks, 1989). Although elm (*Ulmus*),

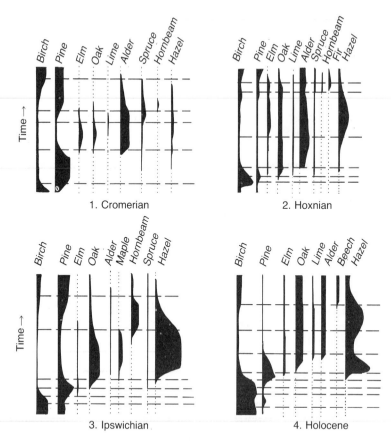

Figure 4.11 Pollen diagrams representing the general sequence of changes in forest composition of eastern England for four different interglacial cycles, numbered 1 through 4 from oldest to youngest. From Wright (1977); reproduced with permission from *The Annual Review of Earth and Planetary Sciences*, Vol. 5, by Annual Reviews Inc.

oak (*Quercus*), and hazel (*Corylus*) arrived simultaneously, hazel was the first temperate tree to increase its populations substantially. Lime (*Tilia*) and alder (*Alnus*) entered the vegetation after the first population expansion and decline of hazel, and beech (*Fagus*) was present only in the late Holocene. The assembly of Holocene forest communities in Great Britain contrasts with that of communities from previous interglacial intervals. In the Ipswichian, the interglacial period prior to the Holocene, which reached its temperature maximum approximately 120 000 years ago, the sequence of replacement of boreal birch–pine forest by temperate forest included the earlier arrival of elm and an early expansion of oak prior to arrival and increase of hazel populations. Alder was much less important than in the Holocene, and beech failed altogether to immigrate across the Straits of Dover. However, maple and hornbeam were well-represented in the early- to mid-Ipswichian, whereas they were not present in substantial relative abundance in the Holocene. In the Hoxnian, the interglacial period preceding the Ipswichian, and in the yet earlier Cromerian, again differences in sequence, timing, and tree species composition are evident from the pollen record (Fig. 4.11). The broad pattern of climatic change alone is insufficient to account for the differences in the timing of arrival and sequence of expansion of the taxa (West, 1961, 1977; M. Davis, 1976; Wright, 1977). Watts (1988) has suggested that long-term changes in the seasonal contrast of temperature and precipitation may alter the relative competitive abilities of plant species. In this context of ongoing climatic influences, species responses to climatic change are individ-ualistic, and additional factors such as seed dispersal, distance from glacial refuge areas, biological factors including competition, and historical factors such as the rate of postglacial rise of sea level that cut off the British Isles from the European mainland all have played important roles in determining the specifics of community assembly in the successive interglacial intervals (West, 1961, 1977; M. Davis, 1976; Wright, 1977; Watts, 1988; Birks, 1989).

In a study of a small lake in the lower Mackenzie River region of the Northwest Territories of northwestern Canada, Jim Ritchie (1985) addressed the question of whether climatic change is the major determinant of vege-tational change, or whether differential migration rates of taxa, competitive interactions among species, or environmental factors, either separately or in combination, can account for periods of relative stability or instability in terrestrial plant communities over the span of an interglacial period. Ritchie (1985) used changes in pollen accumulation rates of a number of individual taxa (Fig. 4.12) as a proxy for estimating trends in their population abun-dances in the region surrounding Twin Tamarack Lake. He found that the 14 000-year record from the site was characterized by two intervals of relative stability in vegetation composition. The first was a tundra community dominated by herbs with dwarf birch (*Betula glandulosa*) that persisted from about 14 000 to 11 000 years ago. The second stable plant community was boreal woodland with alder (*Alnus*), birch (*Betula*), spruce (*Picea*) and willow (*Salix*) from about 5500 years ago to the present. These intervals were separated by a period of instability from 11 000 to 5500 years ago, during which a number of changes in community composition occurred, charac-terized by comparatively rapid change in abundances of constituent taxa. During the period of unstable vegetation composition, there was first a rapid increase in representation of poplar (*Populus*) and a simultaneous decline in herb pollen, followed by peaks in juniper (*Juniperus*) and then willow, after which occurred a rapid decline in poplar and an increase in spruce. During this time interval increases also occurred in pollen accumulation rates for *Myrica*, alder, and grass (Gramineae). Climatic change, with a seasonal maximum in summer warmth reached at 10 000 years ago, was hypothesized

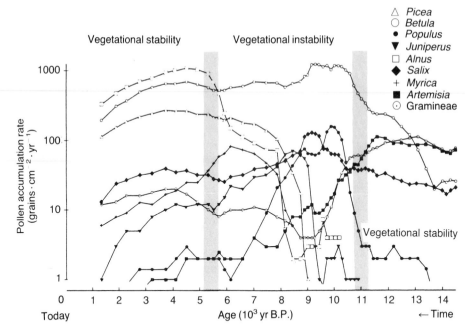

Figure 4.12 Five-sample running means for pollen accumulation rates of plant taxa over the past 14 000 years as recorded in the paleoecological record from Twin Tamarack Lake, Lower Mackenzie Basin, northwest Canada. The vertical bars separate late Pleistocene and late Holocene times of relative vegetational stability from the early-to-middle Holocene time of vegetational instability. From Ritchie (1985).

(Ritchie, 1985) to have initiated the sequence of relatively rapid vegetational change; however, during the period of instability, species interactions were probably more important than climate in determining further changes in community composition. Specifically, the arrival of spruce in the region lagged behind the postglacial climatic change because of the length of time required for its migration northward from source areas to the south (Ritchie and MacDonald, 1986); upon arrival, white spruce (*Picea glauca*) quickly dominated over early-successional poplar and juniper. Expansion of lowland bogs would have facilitated the spread of black spruce (*Picea mariana*). Increases in both alder and tree birch may have resulted from establishment of a disturbance regime that included increased fire frequency (Ritchie, 1985).

The conclusions reached through investigation of the plant fossil record are reinforced by examination of the late Quaternary changes in vertebrate faunas. During the late Pleistocene, mammal communities consisted of species-rich assemblages comprised of species that today are allopatric and

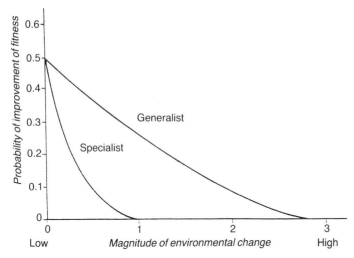

Figure 4.13 Fisher's theoretical model of the probability of increased fitness for both generalist and specialist animals in response to environmental changes of different magnitudes. From Graham and Lundelius (1984).

thus presumably incompatible (Graham and Lundelius, 1984). These faunal assemblages disassembled at the end of the Pleistocene; postglacial changes in community composition were probably initiated by global climatic warming. Graham and Lundelius (1984) argue that because of the great magnitude and rapidity of late-glacial climatic change, changes in faunal communities had a high degree of unpredictability, with the survival of generalists favored over that of specialists (Fig. 4.13). Further, the biota must have responded to environmental changes as individual species, rather than as intact communities of plants and animals. In the latter case, the large herbivores would have been able to track after preferred habitats and migrate with zonally shifting vegetation, rather than becoming allopatric or extinct. However, new community patterns would be expected to emerge given individualistic species responses to global environmental change of the magnitude of the Pleistocene/Holocene transition. For previously coevolved ecosystems, particularly those including large herbivorous mammals (those with adult body size >50 kg; Martin and Klein, 1984), major environmental change would have contributed greatly to their likelihood of extinction. Graham and Lundelius (1984) called this time of community disassembly at the end of the Pleistocene a period of 'coevolutionary disequilibrium'. Sympatric species that survived the extinction event would have redefined niches realized during the Holocene (Graham and Lundelius, 1984). Studies of the modern biogeography of small mammals in the southwestern United States (Lomolino *et al.*, 1989) reiterate the importance of changes in habitat diversity and contiguity since the Pleistocene in affecting both immigration and extinction of species populations and hence development of structure in modern mammal communities in montane environments.

4.4.2 Individualistic species migrations

If species responses to environmental change are individualistic, they will result in differential shifts in distribution. Under these circumstances, in montane environments, elevational shifts in distribution would be expected to result in changing community composition through time, rather than in zonal displacements of plant or animal communities. In a case study from the eastern Grand Canyon in northern Arizona, Ken Cole (1982, 1985) examined changes in distribution of woody perennial plants over the past 24 000 years. This study was based upon the analysis of middens created by packrats (*Neotoma* spp.), which Cole (1982) collected from rock crevices at intervals along an 1800-meter elevational gradient. Plant macrofossils contained in the packrat middens record the local flora, generally within a foraging range of 30 meters of each midden site (Cole, 1985), at the time the midden was constructed. Radiocarbon dates on macroscopic plant material from the middens allowed Cole to array the composition of the samples on both spatial and temporal axes. The plant macrofossils are amenable to specific identification in most cases, yielding a detailed reconstruction of changes in the flora along both elevational and soil moisture (aspect) gradients (Fig. 4.14).

The results of the analysis of changes in vegetation zones in the Grand Canyon enabled Cole (1982) to test the long-held assumption that, although vegetation zones may have been compressed during past times of climatic cooling, the modern parallel between the altitudinal zonation of vegetation and its latitudinal distribution has been a stable pattern through time. On the basis of his set of data, Cole (1982) concluded that Pleistocene vegetation zones in the Grand Canyon were not analogous to modern zones that were depressed altitudinally. Full-glacial assemblages from the Grand Canyon were more similar to modern communities of northern Utah, 450 kilometers to the north. The species have since migrated both altitudinally and latitudinally according to their individual tolerances to changing climates of the Holocene. In the full-glacial period, several species occurred from 600 to 800 meters lower in elevation than they do today, including Utah juniper (*Juniperus osteosperma*), Douglas fir (*Pseudotsuga menziesii*), white fir (*Abies concolor*) and spruce (*Picea* sp.). Limber pine (*Pinus flexilis*) was recorded at least 1000 meters below its modern elevational limit, and shadscale (*Atriplex confertifolia*) was more abundant within a range similar to that occupied today. Pinyon pine (*Pinus edulis*), ponderosa pine (*Pinus ponderosa*), cliffrose (*Cowania mexicana*), manzanita (*Arctostaphylos pungens*), and shrub live oak (*Quercus turbinella*), all found in the Grand Canyon today but not farther north in northern Utah, were, however, apparently absent from the eastern Grand Canyon during full-glacial times. Individualistic responses of species

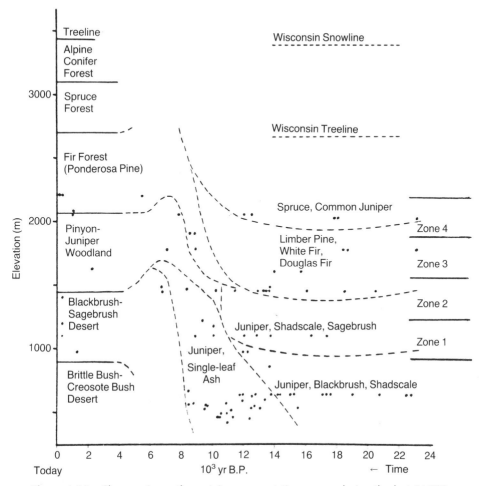

Figure 4.14 Changes in northern Arizona vegetation zones during the last 24 000 years based upon plant–macrofossil data from radiocarbon-dated packrat middens from both the eastern and western Grand Canyon. From Cole (1985).

to changing climate resulted in modern plant communities in the American southwest that are very different in composition from those of the late Pleistocene (Cole, 1982).

Dramatic evidence for individualistic migrations of tree species across continental land masses is provided by the late Pleistocene and Holocene pollen records of both Europe (Huntley and Birks, 1983; Huntley, 1988) and eastern North America (M. Davis, 1976, 1981a, b, 1983; Delcourt and Delcourt, 1987a; Webb, 1988). The methods of compilation of fossil pollen data and of presentation in map form differs slightly among these authors; however, all these studies agree in broad conclusions.

Paul Sears (1942) was the first to suggest the possibility that, for eastern North American forests, during the postglacial interval of northward migrations from southern refuge areas the sequence of arrival of tree species differed across sites located north of the glacial boundary. Sears (1942) was only able to map the relative times of immigration of the taxa to each site, because no means of obtaining an absolute chronology was available at that time. Thus, from the fossil pollen data, it was not possible to map with certainty the directions or rates of tree taxa migrations across the subcontinent. With the advent of radiocarbon dating (Libby, 1955), organic matter preserved in stratigraphic sequences taken from peat bogs and lake sediment cores could be dated. By the 1970s, a large number of paleoecological sites, mostly located north of the late Pleistocene extent of maximum continental glaciation, were radiocarbon-dated and analyzed for changes in fossil pollen assemblages.

A second development that was instrumental in making possible the first mapping of shifts in range limits of individual tree taxa in eastern North America was that of the method of determining absolute rates of pollen accumulation in lake sediments (M. Davis, 1969a, b). Changes in relative percentages of pollen types in a series of fossil pollen assemblages may be misleading if interpreted strictly as changes in the size of populations of the plants that produced them (M. Davis *et al.*, 1973). Because of the percentage constraint, one taxon may appear to increase only because another taxon decreases. Absolute pollen accumulation rates (PAR) are calculated independently for each taxon and are influenced in magnitude by processes affecting their transport and deposition within the sedimentary environment, but not by the relative importance of other taxa (M. Davis *et al.*, 1984).

In a series of papers, Margaret Davis (1976, 1981a, b, 1983, 1986) used radiocarbon-dated sequences of fossil pollen, where possible including those with absolute pollen accumulation rates, to plot the occurrences of major tree taxa both at full-glacial sites south of the glacial margin and first occurrences in postglacial sites farther to the north. She drew contoured maps depicting the positions of northern range limits for each taxon at generally 1000-year intervals during their times of active migrations (Fig. 4.15). The maps depicted the overall directions of migration, which were shown to differ with each taxon in part because of differences in location of their known full-glacial refuges. On these maps, the distances between contour lines for arrival times were used to calculate rates of migration for the tree populations. These migration maps (M. Davis, 1976) were the first to demonstrate conclusively that, in terms of their leading edges of migration, tree species respond to climate change individualistically on a subcontinental scale of resolution.

In order to map changes in the location and dominance of individual tree taxa through the late Pleistocene and Holocene, we adopted yet another approach (Delcourt and Delcourt, 1987a) that involved mapping quantitative reconstructions of forest composition rather than only fossil pollen data. For

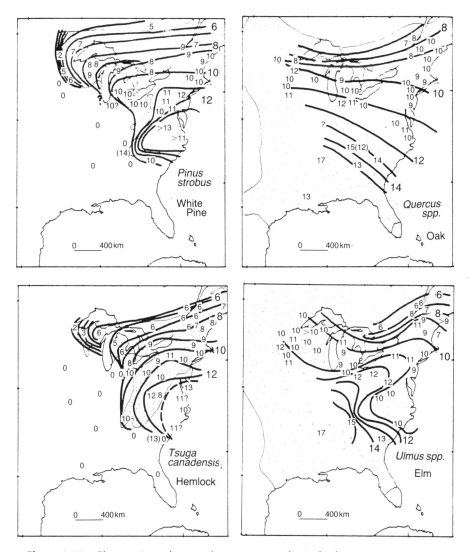

Figure 4.15 Changes in nothern and western range limits for four important eastern North American tree taxa during the late Quaternary based upon radiocarbon-dated pollen records from lake sites. Numbered lines represent range limits at times in the past in thousands of years B.P. The area with stippled pattern represents the modern distributional range of the tree taxa. From Davis (1983).

our study of eastern North American forest history, we assembled a data set of over 1600 paired samples of modern pollen rain and records of contemporary dominance of 19 resolvable tree taxa, based upon the records of the United States and Canadian continuous forest inventories (Delcourt *et al.*, 1984). The pollen-to-vegetation calibrations yielded regression coefficients

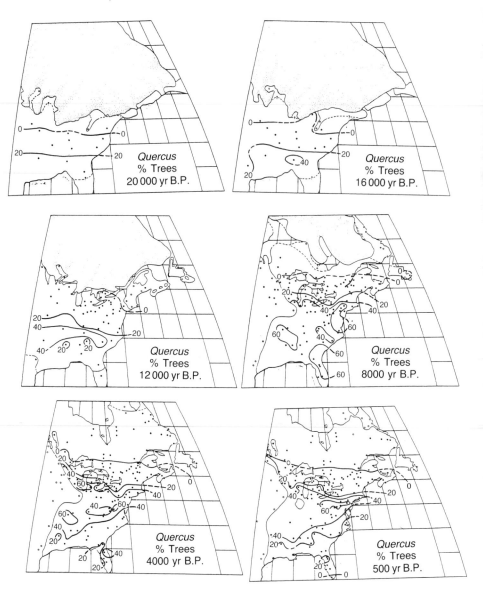

Figure 4.16 Maps depicting changes in range limits as well as population centers, expressed in contoured values of percent dominance in forests, for oak (*Quercus*) trees at 4000-year intervals from 20 000 yr B.P. to the present. From Delcourt and Delcourt (1987a).

based upon data from throughout the modern ranges of the principal tree taxa encountered in the fossil record. When these modern calibrations were applied to the fossil pollen data to calculate changes in past dominance of trees, the results indicated that not only did the rates of migration of the leading edges of species ranges differ, but also the rates of retreating range margins differed, as did changes in the centers of distribution (Fig. 4.16). In addition, the geographical patterns of response of individual taxa to climate change were interpretable in terms of migration strategies that are analogous to life history strategies (Delcourt and Delcourt, 1987a; see the discussion in Chapter 2, section 2.5).

In Europe, pollen-to-vegetation calibrations for individual tree taxa are not generally available on a broad scale because of the extent of long-term alteration of the vegetation by human activities. 'Isopoll' maps depicting the contoured relative abundances of taxa have been used in substitution to give a general indication of changes in distribution and abundance of major plant taxa in the last 13 000 years (Huntley and Birks, 1983). These maps give insight into the locations of full-glacial refuges located south of the Alps and in eastern Europe. Directions of postglacial migration were influenced greatly by the locations of refuges; migration rates were governed largely by different mechanisms for dispersal of seeds. For temperate and boreal regions of Europe, the comprehensive map analyses of Huntley and Birks (1983), Huntley (1988), and Birks (1989) support the individualistic interpretation of species responses to environmental change.

4.5 CONCLUSIONS

1. The position, steepness, and length of environmental gradients and ecoclines are not constant through time and space. Changing environmental gradients that occur with changes in climate affect both the distributions and abundances of plant and animal populations on local, regional, and continental scales.

2. Vegetation change through time can be thought of as a temporal continuum that is analogous to spatial continua on the modern landscape.

3. The hypothesis that biotic communities are tightly coevolved associations that are capable of migrating intact in the face of environmental change is refuted by several lines of evidence from the Quaternary ecological record. Both the plant-fossil evidence and the changes in vertebrate faunal assemblages indicate that climatic change over a time scale of hundreds to thousands of years results in instability of many communities, differential species migration, and changes in structure of communities through time.

4. Triggered by glacial–interglacial cycles of climatic change, disassembly of communities is followed by reassembly that is unpredictable in terms of either species composition or abundance.

5 Factors that structure communities

5.1 ISSUES

In successional time, allogenic factors have been demonstrated to be important in determining the dynamics of community interactions (White, 1979). But to what extent are physical factors important in structuring communities on longer time scales (Barnosky, 1987)? On a Quaternary time scale, what biotic interactions are potentially important in addition to physical factors in structuring communities, and what kinds of interactions occur between biotic and physical systems?

One major goal of many Quaternary scientists is to use the fossil pollen record as a source of information about past changes in climate. Either through evaluation of changes in distribution of indicator species whose modern climatic tolerances are known or, more commonly, by examination of changes in location of assemblages of species today characteristic of particular geographic regions, shifts in climatic variables can be inferred through time and space. Statistical correlations have been made between the present-day positions of major frontal zones or airmass boundaries and vegetation patterns generalized at the level of formation (Bryson, 1966; Bryson and Hare, 1974). These climate–vegetation correlations have been the basis for inferring changes in airmass boundaries through time, based upon paleoecological evidence for general shifts in the positions of physiognomically defined vegetation formations (Bryson and Wendland, 1967; Delcourt and Delcourt, 1981, 1984).

Statistical correlations between pollen assemblages and climate have been used to make quantitative estimates of changes in temperature and precipitation for the Holocene interval (Webb and Bryson, 1972). However, the validity of this method extends in time only so far as the pollen assemblages are quantitatively analogous to modern assemblages (Hutson, 1977). During times of rapid vegetational change or active species migrations, the plant communities producing the observed pollen assemblages change. During the Quaternary, this has led to situations with only poor modern analogues or without modern analogues. Poor-analogue plant communities during, for example, the late-glacial interval may represent communities existing under climatic conditions that are unlike any known today (Overpeck *et al.*, 1985; Delcourt and Delcourt, 1985). Poor-analogue communities may be temporarily out of synchrony with a changing climate, either if climatic change outpaces the rate of dispersal and establishment of the species or if individualistic species migrations cause disassembly of former communities.

Several fundamental assumptions underlie the use of fossil pollen assemblages as proxy data for reconstructing paleoclimates (M. Davis, 1978; Webb, 1980; Birks, 1981b; Prentice, 1986b; Ritchie, 1986). In addition to the assumption about degree of analogue of past pollen assemblages to those existing on modern landscapes, inherent in this approach is the assumption that climate is the 'ultimate ecological control' (Bryson, 1966) on the vegetation that produces the observed pollen assemblages. Another crucial part of this approach is that once a pollen assemblage is documented it can serve as a proxy for a specific vegetation type that in turn has been correlated with climatic variables, thus eliminating the need for an intermediate step in applying correction factors to translate fossil pollen into paleovegetation before inferring paleoclimate (that is, 'pollen calibrated to climate', rather than 'pollen calibrated to vegetation, then vegetation calibrated to climate'). Finally, it is assumed that changes in vegetation and therefore in pollen assemblages do not lag appreciably behind the climatic changes that caused them (Webb, 1980; Huntley *et al.*, 1989).

These assumptions concerning the predominant role of climatic change in producing observed paleovegetational changes cannot be tested readily because the pollen record itself is usually the source for the paleoclimatic record (Birks, 1989). This makes sorting out the different alternative physical and biological factors influencing vegetational change very difficult unless climate, geomorphic processes, soil development, water-level fluctuations, or disturbance regimes such as fire can be documented with quantitative data that are independent of the pollen record (Barnosky, 1987). Once independently derived and given a chronology also determined by absolute dating methods, the several data sets can be analyzed quantitatively in order to distinguish between cause and effect – to detect the existence of potential lags in biotic response to changes in environmental conditions. Interactions of physical and biological factors can then be determined.

The relative importance of leads and lags in Quaternary environmental change and biotic response is a function of the scale at which changes in ecological systems are viewed (M. Davis, 1986; Wright, 1984; Bennett, 1988). If viewed on a very broad spatial scale and with very coarse time resolution, even though individualistic species responses are ultimately responsible for biotic change, vegetation and climate may appear to change in tandem (Webb, 1988). If generalizations are made at the level of major biogeographic provinces or vegetation formations defined on the basis of gross physiognomy and physiological thresholds (e.g., cold hardiness, drought tolerance) defined by generally similar evolutionary adaptations in plant groups (Chabot and Mooney, 1985), then changes in locations of the boundaries of these biomes serve as useful proxies for changes in frequencies of predominant airmasses or other meso-scale climatic variables (Delcourt and Delcourt, 1981, 1984). On a finer scale of resolution, however, the details of behavior of individual species are differential, and other physical and biotic factors ultimately structure communities in addition to climate (Birks, 1981b; M. Davis, 1986; Birks, 1989).

5.2   CLIMATE AND CLIMATIC CHANGE

Climate and climatic change affect the biota differently on different time scales (M. Davis, 1986; Barnosky, 1987; Bennett, 1988). Climate changes on all temporal scales (Bartlein, 1988); relatively short-term changes on the scale of decades to hundreds of years may result in displacements of ecotones, whereas long-term directional changes of 1000-year to 10 000-year duration may effect broad-scale disequilibrium in plant population distributions that can have profound consequences for structuring biotic communities on a continent-wide scale (M. Davis, 1986).

A major controversy among Quaternary ecologists concerns the degree to which the biota is in equilibrium with prevailing climatic conditions and other variables of the physical environment (Prentice, 1986b). An assumption of an equilibrium response of biota to climatic change is essential for reconstructing climate through correlations of climate with present-day distributions of either individual species or assemblages of species (or assemblages of the fossils produced by former plant populations) and for applying either linear or non-linear calibrations to the fossil record in order to reconstruct changes in climate through time (Bartlein *et al.*, 1984, 1986).

Two distinctive approaches have been taken to reconstruct late-Quaternary changes in climate from evidence preserved in the fossil pollen record (Birks, 1981b). The first approach is based upon indicator species. For taxa whose climatic tolerances are known, their presence or absence in fossil assemblages is used to interpret past climatic conditions. The second approach is based upon evaluation of the whole fossil assemblage in terms of the vegetation it represents and the climatic conditions under which that vegetation exists today. In the first case, it is assumed that if the taxon morphology of plant remains preserved in the fossil record stays the same, then no evolution has occurred and therefore no changes in climatic tolerances have taken place through the span of time represented by the late Quaternary interval (Birks and Birks, 1980). This assumption may not be warranted for certain taxa, for example polyploid tundra plants that may have speciated rapidly during deglaciation of the Northern Hemisphere (Bennett, 1988).

The central assumption made in correlating an entire pollen assemblage with existing climatic conditions is that the distribution and composition of those pollen assemblages (and hence the vegetation that produced them) have remained linked in spatial position through time with their corresponding climatic region. A secondary assumption is that these relationships can be estimated quantitatively by linear or non-linear relationships between climatic variables and pollen assemblages. These assumptions both require that modern and past vegetation be in equilibrium with climate and that ecological relationships among species have not changed through time. Thus this technique for reconstructing past climate may not be valid during time intervals of rapid climatic change and vegetational adjustments, during which species migrate at different rates and in different directions across the continents. Existence of fossil pollen assemblages and climatic patterns that lack close modern analogues may invalidate the multivariate statistical approach to climate reconstruction during times of poor-analogue conditions (Birks, 1981b).

The validity of the assumption that the biota remains in equilibrium with climate through time has been questioned by ecologists who consider that not only the timing but also the magnitude of response to a given change in climate are different for each species, with some species responding immediately to annual changes in weather conditions and with others responding only very slowly to long-term directional changes in climate (O. Davis *et al.*, 1986). The result is that over time the structure of communities changes because of non-equilibrium conditions and differential lags in response of species to the same climatic change (M. Davis, 1986). Time lags in species response occur because of inherent differences in life-history characteristics, including life span of individuals, intrinsic rate of increase of populations, and dispersal capabilities of propagules. Inasmuch as climate changes in complex spatial patterns, changes occur in community patterns because these disequilibrium conditions occur both in space and time (M. Davis, 1986).

5.2.1 Tolerance thresholds

One example of a clearly documented relationship between climatic variables and physiological tolerance thresholds of individual species is that of cold hardiness in determining the northern distributional limits of a number of temperate eastern North American trees (Burke *et al.*, 1975, 1976). Many temperate woody plants are capable of withstanding sub-zero temperatures by supercooling water within their tissues, thus preventing the water from crystallizing as ice, rupturing the plant cells, and killing the plant. In a series of experiments on 49 woody plant species, however, it was found that for 25 species of trees indigenous to the temperate zone of eastern North America the supercooling ability reached a common lower limit (Fig. 5.1). All these species were killed by low temperatures that ranged between $-41°C$ and $-47°C$. Comparison of the mapped limits of distribution of the species

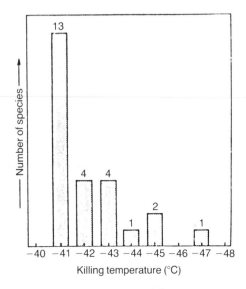

Figure 5.1 Lower cold hardiness thresholds (killing temperatures) summarized for 25 woody angiosperms from the eastern deciduous forest of North America. From Burke *et al.* (1975).

(Little, 1971) with the modal position of the minimum − 41°C winter isotherm (Fig. 5.2) shows a correspondence at the northern margins of their ranges, extending across the region of east-central Canada north of the Great Lakes (Burke *et al.*, 1975). Although the southern limits of distribution for many of the species may be determined by other environmental or biological factors, including nutrient status of soils, hydrological conditions, and inter-specific competition for resources, it is apparent that cold hardiness plays an important role in limiting further northward expansion of the ranges of many species under current climatic conditions.

Past changes in the boundary between boreal coniferous forest and temperate deciduous forest may have been controlled by the position of the minimum − 41°C winter isotherm. Outbreaks of frigid arctic air responsible for severely cold winter temperatures at northern latitudes have characterized the boreal forest region throughout the Holocene interval (Wright, 1981; Kutzbach and Wright, 1985). During the late Pleistocene, however, the physical barrier of the Laurentide Ice Sheet prevented southward penetration of the Arctic Airmass onto the North American continent (Wright, 1981), in part accounting for the equable Pleistocene climate that favored coexistence of faunas composed of species that are today either extinct or allopatric (Graham, 1986). During the Pleistocene, factors other than cold hardiness must be invoked in order to account for past distributions of deciduous

Figure 5.2   Cold hardiness zone map of North America. Region A has average annual minimum temperatures below −40°C and much lower extreme minima. Region B has a finite probability of reaching winter temperatures as low as −40°C. Region C has effectively no probability of winter temperatures reaching down to −40°C. From Burke *et al.* (1975).

forest species (Delcourt *et al.*, 1980). However, for the Holocene, inasmuch as the northern limits of distribution of a large group of tree species are demonstrably controlled by a common, physiologically determined tolerance limit, changes in the boundary between boreal and temperate biomes can be mapped with only minimal complications arising from the individualistic responses of constituent species to changes in other physical or biological factors (Delcourt and Delcourt, 1981, 1987a).

The concept of tolerance thresholds as determining factors in the geographical distribution of plant taxa has been extended to encompass a range of physical environmental factors as predictor variables. In the study of Bartlein *et al.* (1986), species distributions were described as 'response surfaces' whose shape and location are dependent upon variations in a combination of site topography, soil characteristics and disturbance history on a local scale, and of regional trends in macroclimate over a broad scale. Two climatic variables (mean July temperature and annual precipitation) were used to predict the presence or absence of a given pollen type throughout the modern range of each plant taxon investigated (including *Picea*, *Pinus*, *Betula*, *Quercus*, prairie forbs (*Artemisia*, other Asteraceae and Chenopodiaceae/Amaranthaceae), *Tsuga*, *Fagus* and *Carya*). The resulting response surfaces show individualistic variation in the distributions of the taxa with respect to temperature and precipitation variables, for example with prairie forbs showing greatest sensitivity to annual precipitation and with boreal and southeastern conifers (*Picea* and *Pinus*) showing the least sensitivity to this factor. Each of the response surfaces displays one or more optima; in application to the Quaternary fossil record, this would result in more than one possible interpretation of a given abundance value for each taxon and would therefore limit the application of any one set of linear climatic calibrations to a specific region of space and time (Bartlein *et al.*, 1986). Non-linear methods of estimating paleoclimates based on the response surfaces may allow paleoclimate to be estimated if the assumption of vegetational equilibrium with climate is realized (Bartlein *et al.*, 1986; Huntley *et al.*, 1989). Applicability of this technique is also predicted on the assumption that biological factors such as interspecific competition are of only minimal importance in determining the boundaries of a plant taxon's range as well as of the gradients in its abundance across one or more centers of distribution.

5.2.2 Changing biome boundaries

Several studies have demonstrated the statistical correlation between the mean frontal positions of meso-scale airmass boundaries and the modern distributions of biotic communities at the scale of plant formation or biome (Bryson, 1966; Bryson *et al.*, 1970). On this broad subcontinental scale, the correspondence between major features of the global atmospheric circulation system and principal life zone boundaries is striking (Bryson, 1966). For example, the region extending from the Canadian Rockies to the Atlantic Ocean which is today occupied by boreal forest coincides with the area over which continental Polar and Arctic airstreams dominate in winter and where

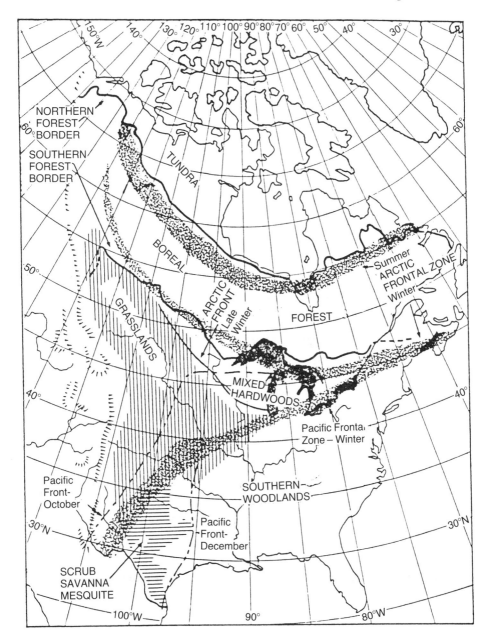

Figure 5.3 Correspondence of mean seasonal positions of airmass boundaries and major vegetation types of eastern North America. From Bryson and Hare (1974).

Pacific and Tropical airmasses predominate during the summer (Fig. 5.3; Bryson, 1966). The tundra climate is defined by year-around dominance of the Arctic Airmass. The prairie region of the Great Plains is characterized by a summer wedge of relatively dry Pacific air; the eastern deciduous forest region is delineated by Pacific and Maritime Tropical airmasses; and the southeastern evergreen forest region is characterized by year-around dominance of the Maritime Tropical Airmass emanating from the Gulf of Mexico. Reid Bryson (1966) considered the present-day correspondence between the distribution of biotic regions of North America and climatic regions as defined by airstreams and airmasses as evidence that the biota is in equilibrium with climate at this scale of resolution. He therefore speculated that post-glacial shifts in biomes determined through analysis of the fossil record could be used to map changes in climatic regions through the Holocene.

Bryson and Wendland (1967) and Bryson *et al.* (1970) applied their correlations of airmass frequency distributions and modern biotic regions to reconstruct broad patterns of change in postglacial climates. They concluded that global climatic changes occurred synchronously and relatively rapidly, with periods between major climate change characterized by quasi-stability. This interpretation was consistent with the climatologists' understanding of the character of general atmospheric circulation, which tends to exist in multiple steady states. They emphasized that while climatic changes would have produced regional ecological changes, the direction and magnitude of biotic change would vary depending upon proximity to major discontinuities (boundaries) in mean airmass frequencies (Bryson *et al.*, 1970).

In a test of the applicability of the Bryson airmass scheme in reconstructing late-Quaternary changes in climate, we mapped changes in distribution of vegetation regions across eastern North America for the past 40 000 years and then inferred glacial and interglacial changes in airmass distributions (Delcourt and Delcourt, 1981, 1984). We compiled paleovegetation maps representing time intervals of relative climatic stability when vegetation was most likely to have been in quasi-equilibrium with climate on a broad regional scale, for example, the full-glacial interval from 20 000 to 16 500 years ago (Delcourt and Delcourt, 1985). Map boundaries were drawn between vegetation types based upon objectively determined percentage values of key pollen types. Despite differential distributions and individualistic responses of plant species during times of rapid climatic change, during the intervals of relative climatic stability, biotic regions at the level of major vegetation types and formations were consistently represented throughout the late Quaternary, providing coherent map units on a broad subcontinental scale of resolution (Delcourt and Delcourt, 1981).

Mapping of changes in airmass boundaries through time (Delcourt and Delcourt, 1984) therefore was generally possible using the criteria of Bryson (1966) for correspondence of climatic boundaries with broad-scale patterns of the biota. The most significant exception was the sparse representation of the tundra biome during the last full-glacial interval, speculatively mapped as a narrow band south of the Laurentide Ice Sheet in eastern North America (Delcourt and Delcourt, 1981). This corresponded with the presumed lack of significant penetration of Arctic Airmass across the North American continent, with frigid temperatures prevalent only in certain glacial reentrants near the border of the Laurentide Ice Sheet (Birks, 1976, 1981a) and at high elevations in the Appalachian Mountain chain (Delcourt and Delcourt, 1984, 1986; Spear, 1989). Reconstructed patterns of distribution of significant discontinuities between airmasses, for example the Polar Frontal Zone whose mean summer position today delineates the transition from boreal to temperate zones, were tested for consistency (Delcourt and Delcourt, 1984) against the marine record of sea surface temperatures based upon fossil foraminifera from sediments of the North Atlantic Ocean (Balsam, 1981). In this comparison of results from independently compiled terrestrial and marine data sets, it was found that shifts in the Polar Frontal Zone were linked through time on a hemispherical scale (Delcourt and Delcourt, 1984).

5.3 CLIMATIC MODULATION OF DISTURBANCE REGIMES

Within each coherent climatic region delimited by airmass boundaries (Bryson, 1966), prevailing climatic conditions influence the character, frequency, and magnitude of dominant perturbations that comprise the disturbance regime (White, 1979). Long-term trends in climate, or even step-wise movement in the mean spatial position of airmass boundaries, result in a changing blend of both type of disturbances and frequency of their typical recurrence intervals. These climatic modulations of disturbance regimes provide ecological forcing functions that influence competitive interactions of plants, disturbance events that dictate the areally prevalent suites of seral stages in plant succession across the landscape, and even thresholds for living tolerances that, when exceeded, selectively eliminate species. Two contemporary contributions in paleoecology document these observations based upon detailed studies in the Great Lakes region of eastern North America.

5.3.1 Frequency of fire occurrence

Konrad Gajewski (1987) explored the importance of physical factors that shape vegetation at the scale of centuries and millennia. Gajewski (1987) examined fine-resolution pollen sequences for the last 2000 years recorded in annually laminated sediments from seven lakes distributed from Minnesota to Maine. These sites were thus representative of the Hemlock–White Pine–Northern Hardwood Forest Region (Braun, 1950), the biotic transition between northern boreal forest and more-southern eastern deciduous forest. Utilizing new paleoecological sites as well as previous studies (Swain, 1973, 1978; Gajewski *et al.*, 1985, 1987), Gajewski (1987) analyzed fossil pollen samples at typically 40-year intervals for the late Holocene; each fossil pollen sample of past vegetation corresponded to the pollen rain preserved for the former time span of about 10 years of lake sediments. Principal components analysis (PCA) was used on the pollen percentage data for each of the seven lake cores, including only the pollen types with a mean value of at least 1% for each site. Gajewski (1987) chose not to focus upon the specific responses of diagnostic or indicator species to climatic change. In contrast to a species-specific or site-specific research design, his strategy was to utilize PCA on an array of sites distributed across one forest region in order to isolate the multiple sources of variation responsible for driving community dynamics across space and through time. The PCA results identified comparable regional trends in all sites, with vegetational dynamics monitored at three levels: (1) over the long term of thousands of years; (2) at medium-frequency oscillations of many centuries; and (3) with high-frequency fluctuations of many decades. Not all vegetational changes were replicated at each of the seven sites. Rather, the timing and general character of biotic adjustment at each lake site was interpreted in the context of the susceptibility of the watershed vegetation to change, based upon its proximity to prominent ecotones, soil moisture and nutrient availability in the nearby landscape soil catena, topographic position, and vegetation state tied to its local disturbance history.

In this way, Gajewski was able to identify broad-scale climate trends tied to increased winter frequency of the Arctic Airmass in the Great Lakes region during episodes of climatic cooling of the late Holocene. Community-level responses were interpreted in the context of the late-Holocene shift of the southern biome boundary of the boreal forest. This involved progressive increases in populations of boreal conifers spruce (*Picea*), fir (*Abies balsamea*) and pine (*Pinus*) and the associated population declines of the more temperate conifer, eastern hemlock (*Tsuga canadensis*) as well as the hardwoods beech (*Fagus grandifolia*) and maple (*Acer*). This southward advance of the boreal forest in the last 2000 years was interpreted to reflect the cold hardiness of boreal species as they tracked the southward displacement of the modal winter position of the Arctic Airmass (Gajewski, 1987).

Gajewski (1987) detected an intermediate level of medium-frequency cycles that occurred over many centuries. Climatic fluctuations of the magnitude and duration of the Little Ice Age (i.e., 1° to 2°C cooling in mean annual temperature from about A.D. 1450 to A.D. 1850 as registered from fossil pollen data in the central Great Lakes (Bernabo, 1981)) produced alternating intervals of relatively cool–wet and warm–dry periods. Corresponding biotic changes were recorded in the time series of PCA scores associated with the second component, with similar times of changes in all lakes from about A.D. 300 to 500, from A.D. 800 to 1000, between A.D. 1400 and 1500, and at about A.D. 1800 (Gajewski, 1987). For example, the second PCA component for Hells Kitchen Lake, northern Wisconsin (Swain, 1978) contrasts the alternating fluctuations in relative abundance for tree populations favored by drier conditions, including pine (*Pinus*) and hemlock (*Tsuga*), with those responding positively to moister conditions, such as maple (*Acer*), oak (*Quercus*), birch (*Betula*), alder (*Alnus*) and elm (*Ulmus*). The cool–wet conditions associated with the Little Ice Age of 350 years ago favored renewed westward migration of beech (*Fagus grandifolia*) and sustained population expansion of mesic trees, particularly in the western Great Lakes region.

Gajewski (1987) asserts that the ecological detail revealed in fossil pollen diagrams with fine-grained temporal resolution permits the ecologist to identify and correctly interpret vegetational dynamics, driven by a variety of physical factors superimposed but occurring at differing frequencies (Fig. 5.4). The application of PCA on time series 'down core' at each site and along a spatial transect of paleoecological sites provides the temporal/spatial array of ecological data necessary to isolate the broad-scale patterns associated with climatic trends lasting over millennia, intermediate-term climatic cycles over centuries, and disturbance events over many decades. At each lake site and its surrounding landscape mosaic, the specific biotic response to climate and disturbance is complex; it may be gradual or abrupt. The impact of physical forcing functions on vegetation may be variously observed (Fig. 5.4): as discrete shifts of ecotones between biomes as biotic thresholds are exceeded; as population fluctuations of indicator species near their distributional limits; or as shifts in successional status across the landscape as the set of disturbed patches experience a change in recurrence interval of disturbance driven by medium-frequency climate cycles. Thus, many frequencies of physical factors are operative and responsible for the ongoing structuring of communities.

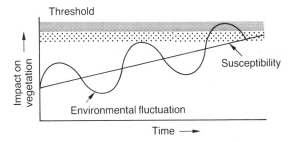

Figure 5.4   Hypothetical model of the susceptibility of vegetation to environmental fluctuations at threshold levels that change with the scale of interaction. From Gajewski (1987).

Gajewski's (1987) study emphasized the necessity of documenting literally hundreds of years of ecological time series to monitor the linkage of climate and disturbance regime as it serves as a highly important control over the suite of landscape-level processes directly shaping the structure of biotic communities. However, Gajewski (1987) left unanswered the key question: what was the means by which climate modulated recurring disturbance events in order to impact population and community dynamics? That is, at the medium-frequency linkage of climate–disturbance–vegetation as identified by Gajewski (1987), how could climatic change requiring centuries directly influence vegetation at a site or across a region?

5.3.2 Periodicities in fire regime

This interaction of climate, environment, and biota has been studied in Lake Itasca State Park, northwestern Minnesota, along the western margin of the Hemlock–White Pine–Northern Hardwood Forest Region. Jock McAndrews (1966) clarified the late-Pleistocene and Holocene history of retreat of glacial ice, and postglacial establishment of major biomes along what is today the steepened ecotone of the prairie-forest border along the 'Itasca Transect' from the western prairie, east through oak savanna and aspen parkland, to mixed conifer–northern hardwoods forest. Contemporary observations of fire ecology are based upon the analysis of fire scars and tree rings, and paleoecological examination of fossil charcoal and pollen assemblages preserved in lake sediments (Frissell, 1973). In this field area, natural wildfire regimes and historic management policies of fire suppression have been characterized. The excellent correspondence of two kinds of independent, detailed local fire histories, based upon fire scars primarily on old-growth trees of red pine as well as charcoal particles contained within annually laminated (varved) lake sediments, has permitted the determination of the frequency (typical recurrence interval between fires on the same site), intensity, and areal extent of wildfire within the watersheds surrounding lakes.

Clark (1988a, b, c; 1989a, b; 1990) examined the impact of climatic change upon fire regimes and their influence on temperate forests. Clark tested the influence of the cool–wet climate of the Little Ice Age on fire regime along the climatically sensitive and fire-maintained ecotones of three biomes along the Itasca Transect. The location of paleoecological sites along the Itasca Transect was chosen to magnify the resolvable impact upon forest communities as registered in plant fossil and charcoal particle stratigraphies from lake sediments. The primary paleoecological site, Deming Lake, was a small (5 ha), deep basin with annually laminated sediments accumulating in as much as 17 meters of water. This kettle-shaped basin formed from the meltout of a block of stagnant glacial ice. Deming Lake is situated within the sandy, well-drained ridge of the Itasca Moraine. Within the lake catchment, forest patches include old-growth red pine (*Pinus resinosa*) and white pine (*Pinus strobus*), with even-aged stands of paper birch (*Betula papyrifera*), large-toothed aspen (*Populus grandidentata*) and trembling aspen (*Populus tremuloides*) occupying tracts burned between 80 and 90 years ago. As old individuals of these aspen–birch stands die, the canopy of the early-successional stand is being broken up and invaded by mesic deciduous species of sugar maple (*Acer saccharum*), basswood (*Tilia americana*) and northern oak (*Quercus borealis*). Using 150 fire scars observed in cross-section slabs of fallen red pine, Clark obtained a local watershed history of fire events spanning the past 400 years. This fire history was independently verified with annual peaks of charcoal production in varved sediments of Deming Lake, and documentation of this charcoal sequence was extended back for a total of 750 years.

Clark (1988a, b, c) refined existing techniques in order to improve the temporal and spatial resolution of fire dynamics as recovered from charcoal particle data. He took segments of lake-sediment core, removed interstitial-pore water with acetone solutions, impregnated the sediment with epoxy resin, and then cut thin slabs (petrographic thin sections) of the varves that were observed by microscope at magnifications of 63× to 125×. The number and cross-sectional area of large charcoal (50 to 10 000 μm long) provided direct measures for the annual production of charcoal particles produced by specific, local fire episodes within the watershed of Deming Lake. These time series of charcoal data permit objective measures of fire frequency (i.e., recurrence time between successive fire events) and the intensity of individual fires, providing a detailed and annual fire history for the past 750 years (annual data for charcoal index in Fig. 3.4; data smoothed with a 15-year running mean in Fig. 5.5). Three intervals of cool–moist conditions were characterized by relatively low ('reduced') fire frequency and moderate amounts of charcoal produced by annual fires, from A.D. 1240 to A.D. 1400,

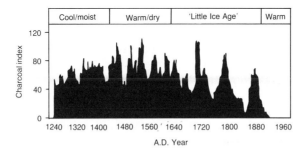

Figure 5.5 Changing fire regimes in northwestern Minnesota reflected in a shift from a 44-year fire cycle in the warm, dry fifteenth and sixteenth centuries A.D. to an 88-year fire cycle after the onset of cooler, moister conditions after about A.D. 1600. From Clark (1988c).

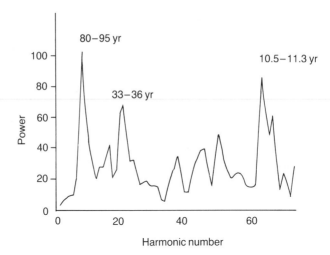

Figure 5.6 Time-series analysis of charcoal accumulation rates within sequences of lake sediments from northwestern Minnesota. The harmonic numbers identified by peaks in the power spectrum correspond to the three primary intervals between regional fire events, that is, fire recurrence intervals of about 11 years, 33 to 36 years, and 80 to 95 years. From Clark (1988c).

as well as from A.D. 1600 to A.D. 1720 and from A.D. 1820 to A.D. 1864 (during the onset, intensification, and the termination of the Little Ice Age, respectively). Increased fire frequency was evidenced by substantial charcoal production (peak accumulation rates generally exceeding 150 cm^2 of charcoal depositing each year on each cm^2 of surface area of lake bottom; charcoal index shown in Fig. 5.5). From A.D. 1920 to present, the management policy of fire suppression coincided with the latest interval during which no charcoal was preserved within Deming Lake sediments.

Time-series analysis of the 750-year sequence of charcoal data revealed four periodicities that accounted for the majority of fire activity (Fig. 5.6; Clark, 1988c; 1989a; 1990). For the full time interval, three harmonics were identified for 11-year cycles, 33-year to 36-year cycles, and 80-year to 95-year cycles. Starting with the initial warm-dry conditions of A.D. 1400, a fourth periodicity of approximately 44 years (43 to 46 years) was prominent during the last 500 years.

These time-series results can be interpreted as evidence for climatic modulation of wildfire occurrence, as constrained by a biological feedback in the form of the accumulation of woody debris as fire fuel load. The two driving functions for this climate–fire–forest interaction are: (1) the strong climatic influence of the 22-year drought cycle that accentuated the regular, pronounced episodes of moisture stress and increased the likelihood of fire (Clark, 1989b); and (2) the time-dependent function of fuel load accumulation following each fire event (Clark, 1989a, b). Following each wildfire disturbance, secondary plant succession generates a cumulative build-up of both fine-textured organic litter and coarse woody debris on the forest floor. The initial scorched earth surface and lack of laterally continuous forest litter tend to limit the spread of minor, ground-surface fires for the first 60 years after a major burn. Early-successional species of birch and aspen readily resprout new saplings from remnant root systems protected below ground from any fire. Given 80 to 90 years, individual trees of the even-aged aspen and birch stands die off, producing a break-up in the canopy, and the assortment of standing boles of dead trees and the fallen snags concentrate fuel load on and near the forest floor. With the replacement of even-aged deciduous forest by uneven-aged mixed conifer–hardwoods, the lateral continuity of flammable matter increases the probability of intense ground and crown fires, especially during drought years characterized by increased incidence of fire.

Within the Hemlock–White Pine–Northern Hardwoods Forest, regenerating stands are particularly vulnerable to wildfire events during episodes of drought stress. The environmental influences of the 22-year drought cycle reflect the increased recurrence of wildfire during years with high drought stress, hence high probability of fire. The time-series analysis of charcoal data revealed the clustering of such fire years with periodic recurrence at approximately 33 years, 44 years, and from 77 to 88 years following the time of successional reset, the last fire on a site (Clark, 1988c). The biological feedback strengthens this climatic modulation of fire disturbance, and intensifies selection for coevolution of birch and aspen in this fire-maintained community.

The paleoecological evidence for past fire regimes demonstrates statistically significant differences in the fire disturbance regime associated with different climatic conditions (Clark, 1989a). Cool–moist climatic conditions favor an interval of 80 to 90 years between fires (Clark, 1988c). With the intensification of the Little Ice Age from A.D. 1600 to A.D. 1864 in northwest Minnesota, fuel load for potential fires accumulated within forests of the western Great Lakes region. Cooler temperatures favored slower rates of decomposition and the slow build-up of moist organic litter. The recurrence interval for wildfire doubled in length, from an initial interval of 36 to 44 years to about 88 years (Fig. 5.5; Clark, 1988c). The less frequent but more intense fires favored the areally extensive spread of individual fires that burned much more of the soil litter, locally impacting the seed bank. One ecological ramification is the shift toward a coarse grain of landscape mosaic during cool–moist times. The widespread fires of the Little Ice Age may have shaped the patterns of tree recruitment and altered the effective success of seedling establishment of different species. The old-growth forests we see today at Itasca State Park reflect initial growth conditions very different from what we see today.

Relatively warm–dry conditions should favor relatively short-term recurrence intervals for the disturbance regime. The 36- to 44-year fire cycles from A.D. 1400 to A.D. 1600 generated more frequent wildfire of low to moderate intensity (Fig. 5.5). The more limited areas of burn would have increased the fine-grain texture of forest types within the landscape mosaic. Paradoxically, fire regimes associated with the warm–dry conditions of the last one hundred years, the time of EuroAmerican settlement and agriculture, have been altered by historic practices of fire suppression, particularly since A.D. 1910 to 1920. The fuel load has had virtually no fire to reduce it, substantially increasing the probability of severely intense, catastrophic fire events unlike any of the natural fires monitored by charcoal time series for any time in the past 750 years (Clark, 1988c). Contemporary patterns of forest disturbance and plant succession may reflect significantly new modes of dynamic processes.

5.3.3 Regional gradients in fire frequency

Serge Payette *et al.* (1989) have recently reinterpreted the significance of wildfire regimes in anchoring the location of ecotones between boreal and arctic biomes within a region of 'virgin' vegetation with no permanent EuroAmerican settlements in northern Quebec. South-to-north gradients in fire regime in the transect of 55° to 59°N (74° to 76°W) coincide with regional gradients in climate (modal summer limit of the Arctic Airmass in Fig. 5.3; Bryson, 1966) and vegetation physiognomy. Along this belt transect from continuous conifer forest to shrub tundra, relative frequency of natural fires in the last 60 to 70 years was comparable across the three biomes: (1) 0.7 fires/year in closed boreal forest; (2) 0.4 to 0.6 fires/year across the subarctic forest-tundra; and (3) 0.4 fires/year in the arctic shrub tundra. However, the more important factor of fire regime was the nearly two orders of magnitude difference in average fire size across this gradient (8000 hectares in boreal forest, decreasing to 5000 hectares in southern forest–tundra, 700 hectares in northern forest–tundra, and 80 hectares in shrub tundra). Integrating these data both on temporal dynamics of fire frequency and area of burns, Payette *et al.* (1989) used the concept of the rotation period of natural fire (Heinselman, 1973) to characterize the typical rate by which fire activities impacted their study region. The rotation period of fire and vegetation regeneration on a short, recurring landscape cycle of only 100 years drives the patchwork mosaic of secondary succession over virtually all of the northern boreal forest. The fire rotation period of vegetation cycling is much longer near treeline, calculated as theoretically greater than 7800 years in the forest–tundra zone, and 9300 years in shrub tundra, time intervals including most of the Holocene interglacial during which this region has been free of glacial ice.

Today, boundaries of the three biomes studied by Payette *et al.* (1989) are, in part, maintained by two physical and biological factors: the ability of fire to spread continuously only in forest stands; and the capability of black spruce (*Picea mariana*), the principal tree species at timberline, to reproduce by seed and to regenerate asexually as krummholz near treeline. On tundra and forest–tundra terrains, the areally extensive cover of lichen produces too little organic fuel to promote the spread of fire. The landscape matrix of lichen cover serves as a natural fire break, protecting isolated spruce clumps from fire. Historic patterns of post-fire regeneration of black spruce confirm that regular seed production and germination consistently occur northward only to the northernmost limit of continuous boreal forest. Episodic recruitment of new spruce cohorts is tied to infrequent growing seasons in which there is successful seed set and germination in the forest–tundra biome. Within the shrub tundra, only isolated clumps of spruce krummholz survive, maintained asexually in the absence of its seed production and germination.

Paleoecological observations of past fire regimes by Payette and Gagnon (1985) indicate that post-fire regeneration of black spruce near treeline is not in equilibrium with contemporary climate. When subjected to fire, some spruce populations in the forest–tundra are unable to recover, and lichen communities replace them. With late-Holocene climatic cooling over the past 3000 years in the Canadian Arctic, fringe populations of black spruce have not been able to reproduce consistently at their northern distributional limits. The resulting expansion of lichen cover has altered the natural fire rotation period. The anastomosing of lichen patches increased the contiguity of effective fire breaks, decreasing the model size of wildfire burn possible at and north of treeline. The insufficient fuel load of lichen communities has limited fire spread and radically altered wildfire regime. Depending upon its location, a discrete wildfire event may lead to the cyclic regeneration of the same plant community, either in the closed boreal forest, or in the open shrub tundra. Within the forest–tundra biome, however, wildfire may trigger permanent replacement of communities, part of the undirectional shift from spruce stand to lichen mat associated with the late-Holocene trend toward climatic cooling, deforestation, and expansion of tundra.

5.3.4 Disturbance thresholds and multiple stable states in vegetation

In the Big Woods area of south-central Minnesota, dominated in presettlement times by elm (*Ulmus*), sugar maple (*Acer saccharum*) and basswood (*Tilia americana*), the combined fossil-pollen study of sediment sequences from small lakes (McAndrews, 1968; Grimm, 1983) and the ecological analysis of 'witness trees' surveyed between A.D. 1847 and 1856 (Grimm, 1984) provides insights into the nature of pattern and process of prehistoric forest dynamics up until the time of initial land settlement by EuroAmerican pioneers (early settlement map of Minnesota vegetation

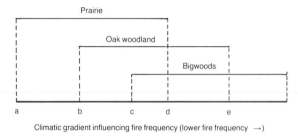

Figure 5.7 Schematic diagram depicting thresholds for establishment of prairie, oak woodland and mixed deciduous forest (bigwoods vegetation) in central Minnesota, plotted along a climatic gradient that influences fire frequency. From Grimm (1983).

prepared by Marschner, 1974). Near the border between prairie and deciduous forest, the character of vegetation change is not observed as a gradual change at a particular site. Rather, as delineated by riverine margins and topographic breaks between gently rolling and hilly terrains, a landscape patch may be occupied by one discrete vegetation type for centuries to thousands of years and then, with passage of a critical threshold, there is an abrupt displacement to another vegetation type (Fig. 5.7; McAndrews, 1968; Grimm, 1983). Grimm (1984) noted that given comparable site conditions of soil drainage and texture, the site history of wildfire events may dictate which of several qualitatively different types of vegetation may persist as apparently stable units through centuries or millennia. Depending upon the nature of the fire disturbance regime, the vegetation may exist in any of 'multiple stable states' as established after a disturbance event (Fig. 5.8). In the Big Woods region, the spatial array of firebreaks is determined by the configuration of stream corridors, wetlands, and lakes on the landscape and by the boundaries between rolling outwash plains and hilly moraines. The fire-probability pattern (Grimm, 1984) of a landscape patch, for example, as governed by its site-specific recurrence interval of wildfire, is influenced by both physical and biotic factors, that in turn constrain the potential vegetation types that can occupy the site. The fire-probability pattern was most substantially influenced by the incidence of fire ignitions by prehistoric Indians, the local distribution and effectiveness of fire breaks, the moisture-holding capacity of soils, particularly those developed on well-drained substrates, the likelihood of fire spread indicated by the areal extent and lateral continuity of the landscape patch and the flammability of the existing vegetation type (based upon accumulation rates of suitable fuel load) (Grimm, 1984). Thus, along a spatial transect crossing from prairie to deciduous forest from southwestern to central Minnesota, landscape patches with very different natural fire probabilities and wildfire regimes produced a mosaic with a variety of successional states following fire events. These included fire-promoted

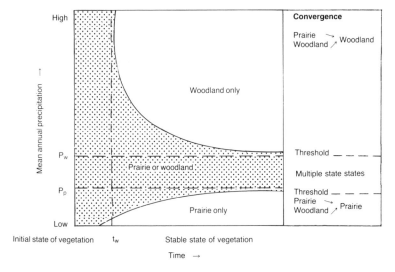

Figure 5.8   Schematic diagram depicting thresholds for establishment of prairie and woodland in central Minnesota plotted on a two-dimensional gradient of climate (influencing fire frequency) and of time. In the upper right quadrant of the diagram (above the woodland threshold for annual precipitation, P_W), only oak woodland will persist as a stable vegetation type once it is established at a site; either of two potentially stable states of vegetation, oak woodland or prairie, can persist in areas with climates bounded by the two climatic thresholds P_W and P_P; and only prairie will persist as a stable vegetation type below the critical threshold for low annual precipitation (P_P) and for increased fire recurrence, conditions represented in the lower right quadrant of the diagram. The symbol t_w represents the minimum time possible for woodland to invade prairie. From Grimm (1983).

communities of prairie, fire-tolerant oak-scrub savanna and aspen communities, fire-infrequent mixed oak forest and fire-intolerant forests composed of elm, sugar maple and basswood forest (designated as bigwoods vegetation in Fig. 5.7).

Grimm (1983, 1984) concluded that multiple stable states characterize the vegetation of the prairie–forest border in the western Great Lakes region. The present-day composition of forest stands cannot be predicted accurately based solely upon knowledge of current physical factors of the environment, because the long-term site history of disturbance and prevailing vegetation state play so powerful a control over these ecological systems.

5.4   SOIL DEVELOPMENT, SUBSTRATE CONTROL AND GEOMORPHIC PROCESSES

In addition to regional climate and climatically influenced disturbance regimes, physical factors on the scale of the local landscape mosaic can be demonstrated to influence long-term vegetational development. These physical factors can be static, as in the case of an extreme, persistent edaphic situation (Brubaker, 1975). Dynamic changes also occur in the physical environment. Paleoecological studies in Europe (Iversen, 1954) and North America (Miller, 1973) document the influence of long-term unidirectional changes in soil profile development, with progressive leaching affecting the nutrient status of the substrate. Fluctuations in soil moisture availability influence the vegetation mosaic on soils with relatively fine texture (Brubaker, 1975). Geomorphic processes including debris avalanches and volcanic mudflows result in vegetational change that may be either sporadic or periodic (Dunwiddie, 1986).

5.4.1 Deglaciation and soil development

Norton Miller (1973, 1980) has used autecological and distributional data on moss species of North America to interpret late-glacial bryophyte assemblages from three plant-fossil sites, including Two Creeks, Wisconsin, Lockport, New York, and Columbia Bridge, Vermont. Radiocarbon dated between 11 400 and 12 100 years ago, these organic-rich deposits of fossil pollen grains, spores, and macrofossils provide evidence for a mosaic of spruce (*Picea mariana* and *Picea glauca*) and tamarack (*Larix laricina*) woodland and more open bryophyte communities, located along the fluctuating shorelines of ancestral Great Lakes. Each of the three plant-fossil sites contains sedimentary deposits associated with late-glacial stages of glacial-ice retreat and meltwater formation of large lakes in front of the melting ice sheet. Because of glacial scouring of nearby limestone and dolomite bedrock, the glacial sediments deposited during ice retreat were initially unweathered and rich in carbonates.

For the species-rich assemblages of predominantly calcicolous mosses represented by late-glacial macrofossils (32 species each at Columbia Bridge and Lockport, and 40 species at Two Creeks), all taxa are still living in North America, and at least 27 species reported for each plant fossil site exist today with at least disjunct populations living within the same state. However, the species distributions are typically more northern, associated with calcareous habitats of boreal or arctic/alpine regions. For the Lockport fossil assemblage, 87% today live in tundra communities across the eastern Canadian Arctic. A modern-day analogue for the late-glacial habitats occurs along the fluctuating shorelines of Lakes Michigan and Huron near the Straits of Mackinac, located at the juncture of the Upper and Lower Peninsulas of Michigan. In this area, calciphile plants colonized beach-pool habitats as well as spring-fed calcareous fens, sites suitable for these plants because of the combination of reduced interspecific competition associated with frequent physical disturbance caused by intense scour of the shore by winter pack ice, as well as the calcareous habitat provided by extensive beaches of carbonate cobbles and sand. Modern moss communities of this region include 93% of those reported in the late-glacial assemblage from north-western New York (Miller, 1973). Along the glacial margin of eastern North America, Miller (1980) described late-Pleistocene environments as poorly-drained, carbonate-rich habitats, with vegetation maintained by disturbance in early- to mid-successional communities. Postglacial soil development resulted in cumulative leaching of carbonates and a gradual shift to neutral or acidic edaphic conditions. As a consequence, populations of calci-phile plants were progressively fragmented, becoming disjunct along their southern distributional limits across the Great Lakes and New England regions. Long-term changes in soil development resulted in the reparti-tioning of habitats, as calciphile plants were successively eliminated from what had been suitable sites in the late-glacial interval. Thus, during the late Pleistocene and the Holocene, long-term change in soils, not climate, was the key factor limiting the distribution of calciphile mosses in the southern portion of their ranges across North America (Miller, 1973, 1980) as well as in Europe (Iversen, 1954).

5.4.2 Vegetation development on contrasting edaphic sites

Linda Brubaker (1975) examined the influence of edaphic factors on development of Holocene vegetation in the Upper Peninsula of Michigan. In this study, she chose small paleoecological sites situated on contrasting substrates with different soil textures, located on a deglaciated landscape. Camp 11 Lake was located on Michigamme glacial till; Lost Lake was situated on Michigamme glacial outwash; and Yellow Dog Pond occurred on Yellow Dog outwash soils (Fig. 5.9). Immediately after deglaciation (10 000 to 8000 years ago in Fig. 5.9), the Yellow Dog outwash plain was occupied by open spruce (*Picea*) woodland, whereas closed forests of spruce and jack pine prevailed on Michigamme outwash and till. Jack pine (*Pinus banksiana*) immigrated onto the Yellow Dog Plains after 9000 years ago, and by 6000 years ago formed open jack pine woodlands that occupied the sandy, dry plains continuously thereafter. Between 8000 and 3000 years ago (Fig. 5.9), white pine (*Pinus strobus*) became important on the relatively drier and more nutrient-deficient soils of the Michigamme outwash plain, with deciduous trees establishing on the finer textured and more nutrient-rich Michigamme till. White pine dominance on the moderately droughty soils of the Mich-igamme outwash during this time interval reflected its competitive advance over hardwoods during the mid-postglacial period of maximum climatic warmth and dryness (Brubaker, 1975). The modern forest mosaic estab-lished after 3000 years ago, with northern hardwoods becoming important across both Michigamme till and outwash, and with jack pine woodlands persisting as an edaphically controlled community on the Yellow Dog outwash plains (Fig. 5.9).

10 000–8000 Years ago

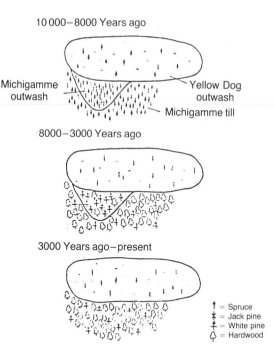

Michigamme outwash

Yellow Dog outwash

Michigamme till

8000–3000 Years ago

3000 Years ago–present

↑ = Spruce
⚹ = Jack pine
⚹ = White pine
Ꙩ = Hardwood

Figure 5.9 Schematic diagram of the patterns of Holocene forest communities distributed across three types of soils, those developed in well-sorted sand of the Yellow Dog Outwash Plains, those in silty sand of the Michigamme Outwash, and those in loamy till of the Michigamme Highlands, located in the Upper Peninsula of Michigan. From Brubaker (1975).

The study of Brubaker (1975) illustrates the role of substrate in influencing long-term vegetation development, depending upon climatic conditions and textural differences of the soil. Soils with very different textures maintain different levels of both soil moisture and nutrients regardless of climatic regime. However, sites with less extreme differences in soil characteristics (e.g., soil formed on Michigamme till and outwash) exhibit marked differences in vegetation only when the level of soil moisture is limiting to plants on one substrate but not the other.

5.4.3 Post-disturbance colonization of lahars

In a study of local plant communities on Mount Rainier, Washington, Peter Dunwiddie (1986, 1987) compared the vegetational development surrounding three ponds developed on the land surface of a 6000-year old lahar, a geological deposit formed by volcanic mud flows and debris flows. Mount Rainier, the largest Quaternary volcano in the Cascade Range, is characterized by a history of geomorphic disturbances during the Holocene that include avalanches, catastrophic deposition of lahars, volcanic eruption and widespread deposition of layers of volcanic ash, and relatively minor re-advances of nearby mountain glaciers. Jay Bath Pond, located at 1311 meters elevation, is located near the boundary between Pacific silver fir (*Abies amabilis*) and mountain hemlock (*Tsuga mertensiana*) forest zones. The paleoecological site of Log Wallow occurs 1.2 kilometers northeast at 1360 meters elevation, 49 meters higher in elevation than Jay Bath Pond; Reflection Pond 1, at 1482 meters, lies 2.4 kilometers east of Log Wallow. For the major Northwestern conifer species, quantitative analysis of needles found as macrofossils in the sediments of the three small ponds yielded a record of changes in representation throughout the 6000-year history of vegetational development around these three sites (Dunwiddie, 1986).

At Jay Bath Pond, about 155 centimeters of organic-rich sediments have accumulated over the lahar deposits; these organic sediments contain layers of volcanic ash throughout the mid- and late-Holocene sequence. Pollen grains and plant macrofossils substantiate that Noble fir (*Abies procera*), Pacific silver fir (*Abies amabilis*) and subalpine fir (*Abies lasiocarpa*) initially colonized the lahar deposit at Jay Bath Pond, along with western hemlock (*Tsuga heterophylla*), western white pine (*Pinus monticola,*) and lodgepole pine (*Pinus contorta*). At Log Wallow, lahar sediments are overlain by about 140 cm of lake sediments including layers of organic detritus and volcanic ash deposited from more recent volcanic eruptions. As at Jay Bath Pond, plant fossils of *Abies amabilis* and *Abies lasiocarpa* were abundantly represented at Log Wallow in sediments representing the time of the initial plant colonization of the 6000-year old lahar. In the 80-cm thick sediment core obtained from Reflection Pond 1, the post-lahar pattern of sediment deposition represented alternating bands of organic-poor and organic-rich layers. Represented by plant macrofossils, Pacific silver fir, mountain hemlock and subalpine fir were the most important tree taxa that colonized the land surface of the lahar deposits.

At all three pond sites, needles of successional species are most abundant for nearly 2500 years after deposition of the lahar. While initial establishment of pioneer plants was probably influenced by the existence of hot, dry conditions on the bare mineral surface of the lahar, the long-term persistence of these species may have been due to warm and dry climatic conditions that reduced moisture availability and encouraged frequent fires (Dunwiddie, 1986). A major fire was recorded by a peak in charcoal fragments within sediments of the three sites just before volcanic deposition of a major layer of volcanic ash about 3400 years ago. Subsequently, with a shift to late-Holocene climatic conditions that were cooler and more moist, forest composition changed to include late-successional trees such as mountain hemlock (*Tsuga mertensiana*) and Alaska yellow cedar (*Chamaecyparis nootkatensis*). The shift in forest composition from early-successional to late-successional species thus was related to three aspects of a dynamically changing environment on Mount Rainier (Dunwiddie, 1986): (1) termination of a significant fire-dominated disturbance regime as reflected in the depositional record of charcoal fragments; (2) influx of nutrients to the watershed from a major ash fall (Mehringer *et al.*, 1977); and (3) a shift to a cool, moist climatic regime that fostered local readvances of nearby mountain glaciers. This study demonstrated the role of geomorphic processes associated with volcanic emplacement of lahars and with deposition of volcanic ash, along with the prevailing disturbance regime of wildfire, in controlling the composition of some montane forests in the Pacific Northwest.

5.5 BIOLOGICAL INTERACTIONS

Changing abundances of plant taxa as revealed by diagrams of changes in absolute pollen accumulation rates (PAR, the rate of deposition of pollen grains on a given surface area of lake sediment per year, a measure of productivity of vegetation; M. Davis, 1963) can, under certain circumstances, be interpreted in terms of biotic interactions, such as competition among plants for light or nutrients on a watershed. During time intervals where no changes occur in climate or in disturbance regimes, competitive interactions among species for light or other limiting factors such as nutrients result in partitioning of realized niches that may be observed as changes in the representation of plant species in the fossil pollen record (Green, 1982; Delcourt and Delcourt, 1987a; Bennett and Lamb, 1988). Mutualistic and parasitic relationships are difficult to document using the fossil record, but several well-known cases of pathogen attacks in Quaternary forest history have parallels with outbreaks of forest pathogens in historic times (Anderson, 1974; M. Davis, 1981b; Patterson and Backman, 1988).

5.5.1 Plant competition

Few cases of plant competition have been directly isolated to explain changes in Quaternary pollen diagrams. Many Quaternary ecologists assume that all observed trends are driven by environmental variables even though independent confirmation of changes in physical environment may be lacking. Also, the prevalent use of relative frequency (percentage) diagrams to display fossil pollen data precludes examining changes in abundances of any one plant taxon in isolation from others. In order to determine the relative roles of physical factors and biotic interactions including interspecific competition in structuring plant communities, three primary kinds of information are necessary (Green, 1982; Delcourt and Delcourt, 1987a; Bennett and Lamb, 1988): (1) time series of quantitative estimates of plant population size or productivity on an individual species basis; (2) independent evidence of any changes in physical environment that are concurrent with the changes in biota; and (3) experimental studies to analyze interactions between pairs of species or among groups of species that are identified by the paleoecological record to have been potential competitors in the past, in order to identify the mechanisms that underlie the observed past behavior of species.

In analyzing the available evidence for past competition in boreal and temperate forest systems of the Western Hemisphere, Bennett and Lamb (1988) considered that the strongest cases can be made for situations in which a species invades a pre-existing community and is followed by a decline in the

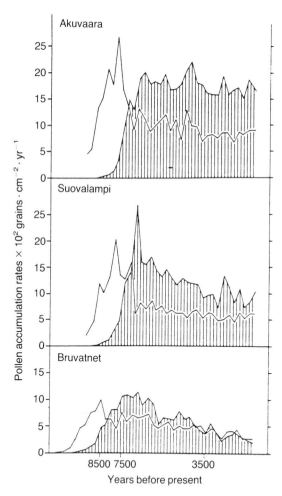

Figure 5.10 Holocene pollen accumulation rates in three lakes in northern Fennoscandia reflecting the successive invasions and subsequent population adjustments of birch (continuous line) and pine (vertical shading). From Bennett and Lamb (1988); redrawn from Hyvärinen (1975).

population level of a species with which it may have competed directly for access to light or nutrients. For example (Fig. 5.10), for several sites in northern Fennoscandia where birch (*Betula*) and pine (*Pinus sylvestris*) today are the only two tree species at or near the treeline, immigration of birch was followed by invasion of pine between 8500 and 7500 years ago. In each case, PAR values for birch declined as those for pine increased, with the amount of birch decline lessening with an increase in latitude. From these data, Bennett and Lamb (1988) suggest that displacement of birch populations by pine occurred as the longer-lived pine species invaded the pre-existing community. Such competitive replacement of shade-intolerant species by shade-tolerant ones has been suggested (Ritchie, 1985; Bennett and Lamb, 1988) to account for restriction of poplar (*Populus*), willow (*Salix*) and juniper (*Juniperus*) populations to localized habitats (and thus their decline in PAR diagrams) upon invasion and rise to dominance of spruce (*Picea*) in north-western Canada between 10 000 and 8000 years ago.

Bennett and Lamb (1988) further speculate that competitive interactions in the past have led to displacements of species along soil-nutrient gradients. For example, partial replacement of hazel (*Corylus*) by lime (*Tilia*) in the woodlands of eastern England about 8000 years ago is inferred as a reduction in the realized niche of hazel because of its displacement from the more acid soils favored by lime. Experimental studies with the two species would help in isolating the factors responsible for their current niche separation.

Another case in which past competitive interactions may help to account for current realized niches of tree species is that of black ash (*Fraxinus nigra*) and ironwood/hornbeam (*Ostrya virginiana/Carpinus caroliniana*) in eastern North America (Delcourt and Delcourt, 1987a). Today, these species all have similar cold-hardiness tolerance limits coinciding with the northern limits of deciduous forest (Burke *et al.*, 1975, 1976). However, black ash is a lowland species, and hornbeam and ironwood are characteristic of mesic to xeric uplands at the modern limits of their distributions. During the late-glacial interval, black ash increased throughout the region of the east-central United States, from Missouri to Kentucky and Tennessee (Delcourt *et al.*, 1986b). After about 14 000 years ago, the fossil-pollen assemblages record a major increase in populations of ironwood/hornbeam throughout that region. Where examined with PAR data at individual sites (references in Delcourt and Delcourt, 1987a), the ironwood/hornbeam expansion was followed by a decline in PAR values of black ash (Fig. 5.11(a)), which conform to a logistic expansion for ironwood/hornbeam and a reduction in black ash of a magnitude consistent with values predicted from a standard two-species model of competitive interaction (Fig. 5.11(c)). During the late-glacial interval, black ash may have occupied a relatively broad portion of its potential or fundamental niche, perhaps enhanced by the equable late-Pleistocene environmental conditions of the early late-glacial interval. We hypothesize (Delcourt and Delcourt, 1987a) that during the time of rapidly

Cupola Pond, Missouri

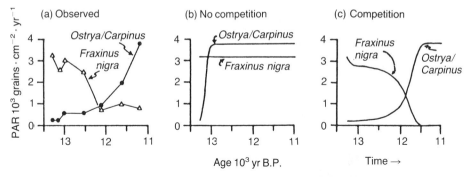

Figure 5.11 (a) Pollen accumulation rates for black ash (*Fraxinus nigra*) and ironwood/hornbeam (*Ostrya/Carpinus*) for the late-glacial interval from Cupola Pond, Missouri. (b) Results of a population growth model of the interactions of two tree taxa without interspecific competition. (c) Results of a population growth model of the interactions of two tree taxa with interspecific competition. From Delcourt and Delcourt (1987a).

changing environmental conditions of the Pleistocene/Holocene transition, ironwood/hornbeam was able to effectively displace black ash to the wettest portion of the environmental gradient, thereby restricting black ash to a more narrow realized niche during the Holocene. As in the case of the relationship of lime and hazel in England (Bennett and Lamb, 1988), this conjecture could be tested with appropriate modern experimental data.

5.5.2 Pathogen outbreaks

One form of biological interaction that has been postulated for the mid-Holocene interval across eastern North America is the outbreak of a forest pathogen that drastically reduced populations of eastern hemlock (*Tsuga canadensis*) throughout its range about 4800 radiocarbon years ago (M. Davis, 1981b). The decline in pollen percentages and influx (PAR) observed for eastern hemlock occurred abruptly over as little as 7 to 50 years at each site where observed. In annually laminated varves preserved in the sediments at Pout Pond, New Hampshire, PAR values of eastern hemlock pollen decreased 75% within a 7- to 8-year period at about 4867 radiocarbon years ago (126.6 cm sediment depth, corresponding in age to 5453 calendar years; Fig. 5.12(b); Allison *et al.*, 1986). A modern analogue for such a widespread and synchronous decline in tree populations is that of the spread of the chestnut blight, a fungal disease (*Endothia parasitica*) introduced from Europe into New England in A.D. 1904 (Anderson, 1974). The rapidity of spread of the chestnut blight and its consequences for virtual elimination of

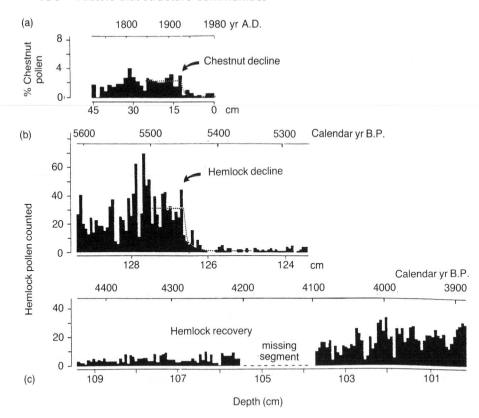

Figure 5.12 (a) Percentages of chestnut (*Castanea dentata*) pollen in sediments of Pout Pond, New Hampshire, during the time interval of regional decline in chestnut tree populations as a consequence of the introduction of the chestnut blight in the early 20th century A.D. (b) Number of pollen grains of eastern hemlock (*Tsuga canadensis*) counted in sediment samples from Pout Pond, New Hampshire, during the time interval of regional decline in eastern hemlock tree populations in the middle Holocene interval. (c) Number of pollen grains of eastern hemlock counted in sediment samples from Pout Pond, New Hampshire, during the time interval of regional recovery of eastern hemlock tree populations in the late Holocene interval. From Allison *et al.* (1986)

mature, host American chestnut trees (*Castenea dentata*) across their natural range over a 50-year period is well documented by historic accounts (Anderson, 1974).

The historic demise of American chestnut is observed in pollen diagrams (e.g. Linsley Pond, Connecticut; Brugam, 1978a, b) as a sharp decline in percentages of *Castanea* pollen and subsequent increases in early-successional trees such as birch (*Betula*) and pine (*Pinus*). For example, at the New Hampshire site of Pout Pond, chestnut pollen values dropped from 3% to <0.5% within a 7- to 8-year period, with the local arrival of the chestnut blight and its infestation of chestnut trees within the watershed (Fig. 5.12(a); Allison *et al.*, 1986). Forest succession following the chestnut blight has resulted in regional differences in modern forest composition, depending upon the availability of forest species (Keever, 1953; McCormick and Platt, 1980; Shugart and West, 1977).

The mid-Holocene decline in eastern hemlock has been attributed potentially to infestation of hemlock looper (*Lamdina fiscellaria*), which may have defoliated its host trees and spread rapidly throughout the range of what may have been a tree species lacking resistance to the insect (M. Davis, 1981b). Alternatively, the insect may have spread from the west where it was parasitic on western hemlock (*Tsuga heterophylla*) but had natural controls in the form of disease or predators that may not have accompanied it as the insect populations spread eastward (Davis, 1981b). Anderson *et al.* (1986) reported the peak in Holocene occurrence of head capsules of microlepidopteran larvae (family Tortricidae, possibly the genus *Choristoneura*) in sediments dating 4800 radiocarbon years ago from Upper South Branch Pond, Maine. This discovery confirmed that the eastern hemlock decline in New England was coincident with the attack of an insect pathogen, which was probably responsible for the severe infestation and middle-Holocene decimation of eastern hemlock populations. Patterson and Backman (1988), however, caution that an outbreak of the spruce budworm (*Choristoneura fumiferana*) may have decimated local populations of a primary host such as balsam fir (*Abies balsamea*) and that the decline in eastern hemlock populations may have been unrelated to this particular episode of insect infestation 4800 years ago.

Recovery of eastern hemlock populations after the decline at about 4800 radiocarbon years ago continued over a span of time as much as 2000 years (Fig. 5.12(c); Allison *et al.*, 1986). This situation is an example of a 'natural experiment of the past', in which selective removal of a structural dominant occurred without catastrophic disturbance to the rest of the forest by windthrow, by fire, or by a shift in climatic regime. Forest succession took a different course in each geographic region thus far investigated (M. Davis, 1981b).

In New Hampshire, the eastern hemlock decline was followed by increases first in pollen accumulation rates (PAR) of birch (*Betula*), then beech (*Fagus grandifolia*), sugar maple (*Acer saccharum*) and hornbeam (*Ostrya virginiana* or *Carpinus caroliniana*). Eastern hemlock exhibited resilience in this portion of its range, recovering its former abundance after about 2000 years (M. Davis, 1981b), a length of time representing approximately 20% of the Holocene interglacial interval. This extended period of recovery may reflect the time necessary for coevolution of eastern hemlock and its pathogen, either by development of resistance to the pathogen by eastern hemlock or by increased biological control of the pathogen by disease or predators (M. Davis, 1981b).

In contrast, in the Upper Peninsula of Michigan (Brubaker, 1975) eastern hemlock was replaced by birch (*Betula*), then by fir (*Abies balsamea*), spruce (*Picea*) and pine (*Pinus*), collectively forming a new forest type that characterized the late Holocene. In the southern portion of its range in the southern Appalachian Mountains of southwestern Virginia, eastern hemlock was succeeded by oak (*Quercus*) and hickory (*Carya*) and, as in Michigan, never regained its former abundance (Delcourt and Delcourt, 1986).

If the pathogen attack on eastern hemlock about 4800 radiocarbon years ago was a density-dependent phenomenon, it would have been expected to have the greatest impact in the population centers of eastern hemlock. Maps and histograms showing changes in distributional range as well as in population size of eastern hemlock through the mid- and late Holocene intervals (Delcourt and Delcourt, 1987a) are consistent with this speculation. Through the mid-Holocene interval, eastern hemlock continued to extend its range to the north and northwest across the Great Lakes region even during and after the major reduction in its population centers 4800 years ago. The eastern hemlock decline also may have been a cyclic phenomenon analogous to the cyclic infestations of spruce budworm (*Choristoneura fumiferana*) today across the Great Lakes and eastern Canada regions (Anderson *et al.*, 1986). Paleoecological studies from early-Holocene population centers of eastern hemlock in the southern Appalachians record progressively increasing fluctuations of hemlock populations from 9000 years ago to 4800 years ago (Delcourt and Delcourt, 1986).

Eastern hemlock was not driven to extinction by the major mid-Holocene attack of an insect pathogen. However, in much of its range, the competitive relationships of eastern hemlock with other temperate forest trees was altered sufficiently by this biological event of hemlock decline that subsequent forest recovery resulted in regional differentiation of new forest communities (M. Davis, 1981b; Delcourt and Delcourt, 1986, 1987a).

5.6 CONCLUSIONS

1. The Quaternary ecological record demonstrates that both physical and biological factors are important in structuring biological communities.

2. Climate and climatic change is a significant factor in driving ecological processes that is operative over a broad spectrum of temporal and spatial scales. Climate modulates disturbance regimes that, in turn, dictate the landscape mosaic of patches occupied by successional communities.

3. The combination of frequency of disturbance recurrence and its areal extent of impact on biotic communities is integrated in a natural rotation period for a given disturbance regime.

4. The prehistoric trajectory of vegetation development and the site-specific history of disturbance directly influence the likelihood of persistence

for one or multiple potentially stable states of vegetation. The dynamic nature of community change may be gradual, or it may be abrupt with passage of environmental or biotic thresholds.

5. Biological interactions involving competition and pathogen outbreaks have important ecological consequences on a Quaternary time scale.

6 Ecosystem patterns and processes

6.1 ISSUES

Attributes of ecosystems that are relevant to understanding their stability through time are amenable to both theoretical analysis and field testing; these ecosystem attributes include resilience, or how fast species composition returns to equilibrium following a perturbation such as a reduction in species, and persistence, the length of time a particular type of ecosystem endures (Pimm, 1984). The relationship of persistence of species composition to species richness is dependent upon the time scale of observation (Pimm, 1984). The Quaternary ecological record can be used to test questions of ecosystem stability by examining evidence for changes in resilience and persistence of ecosystems through time. Changes in these ecosystem attributes may be related to changing thresholds of nutrient availability in soils and aquatic environments, changing species composition as a result of postglacial immigration from refuge areas (M. Davis, 1976, 1986), and anthropogenic activities that may reduce numbers of native species, cause changes in nutrient status of soils, and alter trophic relationships through selective removal of predators or herbivores (Binford et al., 1987). Further, the paleoecological record of ecosystem changes over one to many glacial/interglacial cycles offers insight as to whether ecosystem development through time is unidirectional or is cyclic, and helps to identify underlying mechanisms for ecosystem changes (Birks, 1986).

6.2 CYCLES OF ECOSYSTEM DEVELOPMENT

Based primarily on the last (Wisconsinan or Weichselian) glacial episode and on the subsequent Holocene interglacial interval, overall changes in ecosystem properties through time have been proposed to follow a more or less predictable sequence or cycle of events (Iversen, 1958; Birks, 1986). The glacial–interglacial cycle of events occurs on the level of whole ecosystem interactions, including interrelated changes in vegetation, fauna, and soil fertility on watersheds, and nutrient and trophic status of lakes. As a paradigm of vegetation development across northern Europe, the glacial–interglacial cycle proposed by Iversen (1958) and later modified by Andersen (1966) and Birks (1986) consists of four principal phases characterized by major changes in climate, soils, and plant communities (Figure 6.1(a)).

The initial, glacial-age phase (cryocratic phase *sensu* Iversen, 1958) is also the longest in duration, representing the majority of each glacial–interglacial cycle which, in Europe, was cold, dry, and continental in climate (Behre, 1989). In North America, climates were colder than they are today but more equable, with lesser seasonal extremes than were experienced across the European continent (COHMAP, 1988). In both cases, the biota immediately south of the continental ice masses consisted of mixtures of plant and animal species unlike any represented today; in Europe, arctic–alpine, steppe, and ruderal herbs formed sparse vegetation cover on base-rich mineral soils, continually disturbed by frost heave and the alternate freezing and thawing of soil moisture (Birks, 1986; Behre, 1989). In temperate latitudes of North America, large Pleistocene mammals co-existed with arctic, boreal, and warm–temperate small mammals. The full-glacial vegetation mosaic included both forests of boreal conifers and some temperate deciduous species, but arctic tundra vegetation was restricted to very small pockets along the southern glacial-ice margin (Graham, 1976; Graham and Lundelius, 1984; Delcourt *et al.*, 1980; Delcourt and Delcourt, 1981; Wright, 1981). Birks (1986) characterized this 'cryocratic' phase as an extended period of apparent vegetational stability, on the basis of fossil pollen assemblages that were relatively consistent through both time and space across large portions of Europe. Birks (1986) speculated that because cold climatic conditions characterize up to 90% of Quaternary time, long-term ecosystem stability may be the predominant biotic pattern on this time scale.

On continents of the Northern Hemisphere, the onset of interglacial conditions (the 'late-glacial' interval) is marked by a relatively rapid increase in temperature, recession of ice sheets, and a changeover in landscapes from glacial to periglacial regimes or from periglacial to fluvial geologic processes. In Europe, this 'protocratic' phase (Iversen, 1958) is characterized by immigration and establishment of pioneer communities on newly available, unleached alkaline and fertile soils and in base-rich lakes that support a pioneer aquatic flora (Birks, 1986). Across mid- to high latitudes of both Europe and North America, this late-glacial interval is a time of individualistic migration of species, rapid biotic change, vegetational instability, and formation of ephemeral communities with poor modern analogues (M. Davis, 1983; Birks, 1986; Delcourt and Delcourt, 1987a; Jacobson *et al.*, 1987; Huntley, 1988). Overall, environmental controls are the predominant forcing functions for biotic change during this part of the glacial-interglacial cycle of vegetation development. The biota is in disequilibrium with climate over large areas, and the high rate of floristic turnover results in maximal species and community diversity at the landscape scale (Delcourt and

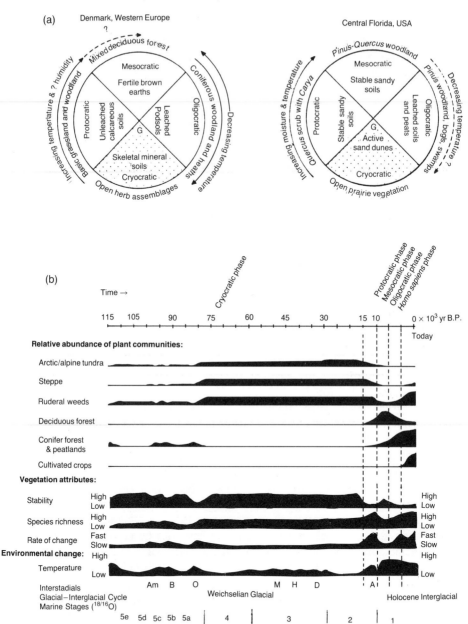

Figure 6.1 (a) The interglacial cycle in Denmark and central Florida, showing four major phases of changes observed in regional vegetation, soils, and climate (probable climatic changes are denoted by a solid line; unproven climatic changes by a broken line with a ?). (b) Changes in vegetation during the past 115 000 years of the last glacial–interglacial cycle in northwest Europe. Am = Amersfoort; B = Brørup; O = Odderade; M = Moershoofd; H = Hengelo; D = Denekamp; A = Allerød. Modified from Birks (1986).

Delcourt, 1983; Birks, 1986). This is reflected in the wide array of late-Pleistocene mammal communities, with fossil animal assemblages for which there are poor or no modern analogues (Graham and Lundelius, 1984; Semken, 1983; Graham, 1976). The protocratic phase is a time of biological crisis characterized both by extinctions of floral and faunal elements and possibly by speciation events (Birks, 1986; Delcourt and Delcourt, 1983, 1987a).

During the 'mesocratic' phase of interglacial ecosystem development (Iversen, 1958), mid-latitudes of both northwest Europe and eastern North America develop temperate deciduous forests with fertile, brown-earth soils; lakes become fertile with a diverse biota (Kapp, 1977; Birks, 1986). Climate is as warm or warmer than during the previous, protocratic phase, but migrational lags and other disequilibrium effects are less important as biological interactions become more important. Vegetational changes observed in pollen diagrams during the mesocratic phase include the arrival, establishment, and population expansion of mesic deciduous trees (Watts, 1973; Bennett, 1983) followed by the decline and readjustment of previously established shade-intolerant tree populations because of interspecific competition for light and nutrients (Bennett and Lamb, 1988; Delcourt and Delcourt, 1987a). Readjustments in realized niche breadth result for both plants (Bennett and Lamb, 1988; Delcourt and Delcourt, 1987a) and animals (Graham and Lundelius, 1984). The specific composition of mesocratic, terrestrial, vascular plant communities varies from one interglacial interval to the next because of historical factors that influence the specific sequence and timing of re-immigration of taxa (West, 1961, 1977). Mesocratic ecosystems that develop in successive interglacial intervals, however, are broadly similar in physiognomy of vegetation and in nutrient status of soils, and may be considered examples of multiple stable states in ecosystem development (Birks, 1986). In the lower Midwestern United States, establishment of both upland prairie and temperate deciduous forest across the region described as the 'Prairie Peninsula' (Transeau, 1935; Wright, 1968) may constitute two metastable states (O'Neill *et al.*, 1986), each representing a separate trajectory of biotic change diverging past a 'bifurcation point' of postglacial vegetation development. In part because of changes in forest composition, but primarily because of progressive leaching of nutrients from the soils, the nutrient inputs to lakes during the mesocratic phase result in base-rich and productive water bodies in which organic-rich sediments accumulate (Birks, 1986). Aquatic plant assemblages include floating-leaved macrophytes as well as submerged aquatic plants (H. H. Birks, 1980).

Several major factors are involved in the change from the mid-interglacial mesocratic phase of ecosystem development to the late-interglacial 'oligocratic' phase (Andersen, 1966). The first of these factors is a change in climate towards an overall decrease in mean annual temperature that tends to favor coniferous trees over deciduous trees in temperate latitudes (Iversen,

1958). The second factor is the tendency toward progressive leaching of nutrients from forest soils (Andersen, 1966). Particularly in the humid interglacial climate of northwest Europe, continued leaching of soils results in podzolization, with progressive nutrient loss, soil acidification, and eventual soil degradation as the soil fauna changes from earthworms to arthropods and eventually becomes depauperate as soils change from a mull humus to a mor (Andersen, 1984; Stockmarr, 1975). With such deterioration in the fertility of soils, European deciduous forests decline during late-interglacial times, stressed because of changing edaphic conditions, and they are replaced by coniferous forests, acid blanket bogs, or heaths. A third influential factor in late-Holocene times that has enhanced this tendency toward ecosystem change is forest clearance, agriculture, and other land use by humans (the *Homo sapiens* phase of Birks, 1986, Figure 6.1(b)). In Europe, this anthropogenic phase of impact on vegetation development began in Neolithic times, between 6000 and 4500 years ago (Barker, 1985). In the British Isles, the combination of soil podzolization and forest clearance led to increased hardpan development, decreased soil aeration, high groundwater levels, and paludification over extensive areas, resulting in widespread development of blanket mires (Moore, 1975, 1984). In southeastern North America, Watts (1980a) speculated that the mid- to late-Holocene changeover from oak-hickory forest to pine forest may have resulted from the cumulative effects of long-term nutrient leaching and soil deterioration (Fig. 6.1(a)). Delcourt (1980) suggested as an alternative hypothesis that the changeover from deciduous to coniferous forest in the Southeast may have occurred because of fires set by Archaic Indians for driving game and encouraging plant growth used as browse by herbivores.

Watts (1988) proposed a synthetic framework for reinterpreting Iversen's model (Iversen, 1958) for cyclic glacial–interglacial changes in climate and their influence on long-term development of vegetation and soils. This framework is based on the widely adopted paradigm that Quaternary climatic changes are caused by the changing astronomical configurations of the Earth and the Sun (Imbrie and Imbrie, 1979). The combination of systematic changes in the Earth's orbit about the Sun and changes in the wobble of the Earth about its axis of rotation result in changes through time in both the amount and seasonal contrast of solar radiation received by the land surface. These orbital changes drive climatic changes in three regular 'Milankovitch cycles' of approximately 21 000 years, 41 000 years, and 100 000 years (Imbrie and Imbrie, 1979). As a consequence, the Earth's climate is continually changing and biological populations are potentially never in equilibrium with climate; rather, the biota is always tracking an ever-changing climate (Bartlein, 1988). During the last 700 000 years of the Quaternary, however, the 100 000-year Milankovitch cycle has predominated, resulting in a step-wise oscillation between two fundamentally different climatic regimes. During Ice-Age conditions of cold, glacial climates, continental ice sheets expand over a 90 000-year interval. Glacial-age conditions are terminated abruptly, followed by warm interglacial conditions that typically last about 10 000 years, during which much of the ice stored in continental ice sheets melts. Evidence for twenty glacial–interglacial cycles during the last two million years is derived from analysis of the ratios of oxygen isotopes in the shells of fossil foraminifera deposited in continuous deep-sea sediments (Imbrie and Imbrie, 1979). The change from glacial to interglacial conditions is initiated when the amount of sunlight received by the Earth is greatest and the seasonal contrast is most extreme at middle and high latitudes in the Northern Hemisphere. This time of enhanced seasonality results in very hot summers and very cold winters (Fig. 6.2; Kutzbach and Guetter, 1986). The climatic transition from interglacial to glacial mode occurs when both the mean global temperature and the seasonal contrast between summer and winter are reduced (Imbrie and Imbrie, 1979).

The Iversen model (1958) of glacial–interglacial vegetation cycles emphasized three environmental and vegetational changes: the progressive warming of average temperatures from glacial times to an interglacial peak and then gradual cooling associated with the onset of the next glacial episode; the availability of fresh, unweathered substrates on deglaciated land that were

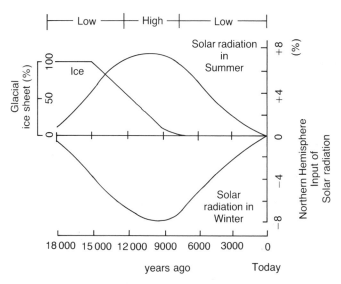

Figure 6.2 Schematic representation of the changes in the Northern Hemisphere climate during the past 18 000 years with systematic changes in both seasonal contrast of solar radiation and extent of glacial ice through the last glacial–interglacial cycle. Modified from Kutzbach and Guetter (1986).

exposed to long-term leaching of nutrients and a gradual change in soil pH from alkaline to acidic conditions; and the change in vegetation from dominance by shade-intolerant plants to that of shade-tolerant plants as the local environment became more buffered beneath a closed forest canopy. Watts (1988) proposed a major modification of this scheme that instead emphasized the overriding effect of climatic control of the biota, particularly as it is related to the changing seasonal contrasts between summer and winter as driven by Milankovitch cycles. Using the seasonal range in temperatures for the Holocene as an explanative model for earlier interglacial intervals, Watts (1988) suggested that the onset of each interglacial was characterized by an anomalous interval of extreme seasonal contrasts, such as occurred between about 13 000 and 8000 years ago (Fig. 6.2). This would result in an extension in length of the growing season and the availability of new, unoccupied habitats in deglaciated terrains that would be invaded by pioneer tundra and then boreal forest communities. In Europe, the peak in climatic seasonality at about 10 000 years ago would have favored the establishment of early-interglacial temperate forests dominated by deciduous tree species of oak, elm and lime, which previously had been limited along their northern distributional ranges by low spring and summer temperatures. Another suite of trees, including European beech, that had been limited by cold winter temperatures during the time of maximum seasonality may have expanded in range and became important in the vegetation relatively late in the interglacial cycle, when both mean winter and summer temperatures became more equable (Figs 4.11 and 6.2). Thus, the postglacial migrations and range expansions of plant taxa must be viewed in terms of their climatic thresholds of tolerance during particular seasons of the year. The recurring pattern of interglacial vegetation development can be interpreted in terms of individualistic species responses to Milankovitch forcing of seasonality of climate within the context of the life-history and migrational strategies of the species, as well as to other long-term environmental changes in the sites into which the plants have invaded (Watts, 1988).

When examined at the ecosystem level, it is clear that glacial–interglacial cycles of climatic and environmental change evoke cyclic changes in the biota. The rates, directions, and timing of ecosystem changes are governed both by extrinsic factors such as changes in geomorphic and other disturbance regimes and progressive soil leaching as well as by intrinsic factors including intraspecific and interspecific competition. Threshold effects can be observed, for example in the process of soil retrogression, that determine the status of the entire ecosystem. Ecosystem stability, in terms of persistence and resilience to perturbations, can be demonstrated to be determined by long-term changes in environmental thresholds as well as changes in species number as extinctions occur in large herbivores and top predators and as structural dominants such as deciduous trees immigrate into previously established communities. Over the last glacial–interglacial cycle, the

'cryocratic' glacial interval has been characterized by relatively stable, persistent ecosystems that in some cases included a mix of floristic and faunal
elements existing under a range of soil conditions and disturbance regimes
unlike those of any modern ecosystems (Guthrie, 1982). These cryocratic
ecosystems were resilient under the set of perturbations that characterized
90% of the Quaternary interval (Birks, 1986; Fig. 6.1(b)). However, with the
onset of rapidly changing environmental conditions during the late-glacial,
protocratic phase, ecosystems were destabilized both by loss of species in
many trophic levels and by creation of newly available territory with fresh,
unleached soils for colonization by pioneer communities. Subsequent
changes in ecosystems were rapid and progressive with diminished resilience
to a constantly changing physical and biotic context that eventually included
humans as a major factor.

6.3   HUBBARD BROOK/MIRROR LAKE WATERSHED

One of the primary sites in North America for long-term ecological research
and monitoring is the combined Hubbard Brook Experimental Forest and
Mirror Lake Watershed, designated a Long-Term Ecological Research
(LTER) site in the New England region. In this ecosystem study area of
northern New England, an experimental approach for whole ecosystems has
been used to contrast the response of natural forested and manipulated
watersheds to different disturbance regimes and management practices
(Bormann and Likens, 1979; Likens *et al.*, 1977). Bormann and Likens
(1979) applied the paradigm of steady-state dynamics (Fig. 6.3) to the
forested watershed at Hubbard Brook, envisioning a shifting patchwork of
successional stages of vegetation following natural disturbance events across
the landscape. This steady state would constitute the long-term persistence
of a full mosaic of vegetational communities, collectively maintained
somewhere within the landscape, although each vegetation type would
occupy different sites at different times. According to the model of Bormann
and Likens (1979; Fig. 6.3), an initial disturbance such as clear-cutting
would lead to a reorganization phase during which total biomass declines,
run-off increases, and loss of both particulate matter and dissolved nutrients
is accelerated. Following this reorganization phase, an aggradation phase is
characterized by a cumulative increase in total biomass to peak levels, and
increased biological regulation of nutrients, with decreased run-off, decreased rates of erosion, and increased sequestering of dissolved nutrients.
During the transition phase (Fig. 6.3), biomass levels decrease as nutrient
export in stream water increases. Finally, during the shifting-mosaic steady
state phase, total biomass stabilizes, with minor fluctuations about a mean
that is regulated by long-term tightening of biogeochemical cycles with
stabilization of the hydrology in terms of overland run-off and export of

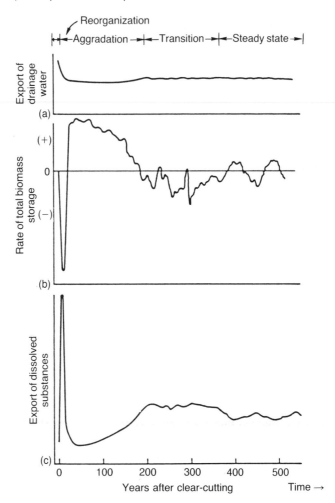

Figure 6.3 Steady-state model of long-term responses of a temperate hardwoods forest ecosystem for a 500 year interval following clear-cutting of the forest: (a) drainage water export; (b) rate of biomass storage (+ = ecosystem accumulating biomass, − = losing biomass); (c) export of dissolved substances in drainage water. From Bormann and Likens (1979).

particulate matter and nutrients. The shifting-mosaic steady state hypothesis (Bormann and Likens, 1979) allows for long-term shifts in mean values as a function of changes in weathering rates of substrate and soil development, but it does not explicitly take into consideration the possibility of major climatic change (M. Davis, 1985).

The paleoecological record spanning the last 14 000 years from Mirror Lake provides a direct test of the shifting-mosaic steady state hypothesis for

New England forest dynamics (M. Davis, 1985). The ecological research conducted at the Hubbard Brook/Mirror Lake LTER site provides a characterization of the linkages between terrestrial and aquatic ecosystems (Likens and Bormann, 1974). The sedimentary, geochemical, and biological records provide long time series of observations of ecosystem-level changes within Mirror Lake and its surrounding watershed (Likens and Davis, 1975; Davis, 1985; Davis *et al.*, 1985a, b). These sediments reveal an intricate history of the postglacial evolution of the watershed and its soils, the initial establishment and subsequent changes in terrestrial vegetation, and changes in water quality driving trends in lake productivity. The export of water and water-soluble nutrients, mineral particles, and organic debris from the surrounding watershed to Mirror Lake represents the principal, gravity-driven linkage between terrestrial and aquatic systems. The sedimentary deposits of inputs to Mirror Lake reveal the answers to the following ecological questions. (1) How are these ecosystems impacted by changes in climate, geomorphic processes, soil development, and the invasion sequence of species migrating into the study area? (2) Are these ecological systems persistent in steady state, or are they vulnerable to change? (3) Does a change in nutrient retention and recycling in upland communities trigger a cascading shift in water quality and in lake trophic status?

Mirror Lake (43°56.5′ N, 71°41.5′ W) is situated at an elevation of 213 meters in the foothills of the White Mountains in northern New Hampshire. Mirror Lake drains into Hubbard Brook in the lower part of its valley. In contrast, the experimental watersheds are situated near the headwaters in the upper portion of the Hubbard Brook Valley. Mirror Lake is small (15 hectares), oligotrophic, nutrient-poor, and slightly acidic (pH 5.5–6.5), with a maximum water depth of 11 meters. The watershed is 103 hectares, with a layer of surficial glacial deposits of sandy drift up to 30 meters thick mantling Devonian-age crystalline bedrock of slate, schist, and quartz monzonite (Winter, 1985). Between 1823 A.D. and the early 1900s, several dams were built on the lake outlet; this historic rise in water level by 1 to 2 meters provided a holding basin to store logs as well as water power to run sawmills (Likens, 1985). Within the last two centuries, EuroAmerican settlement within the watershed resulted in forest clearance, then forest regeneration. Modern forests are dominated by yellow birch (*Betula allegheniensis*), sugar maple (*Acer saccharum*) and beech (*Fagus grandifolia*), with more restricted stands of eastern hemlock (*Tsuga canadensis*), oak (*Quercus*) and pine (*Pinus*).

Mirror Lake is a kettle lake, with its conically-shaped basin formed about 14 000 years ago by the late-glacial melting of a stagnant block of glacier ice buried in glacial-till deposits of a moraine (Winter, 1985; Fig. 6.4(a)). Maximum sediment accumulation of up to 13 meters extends across the central portion of the lake basin (M. Davis *et al.*, 1984; Fig. 6.4(b)). Progressive filling of the conical basin with lake sediment occurred in a spatial

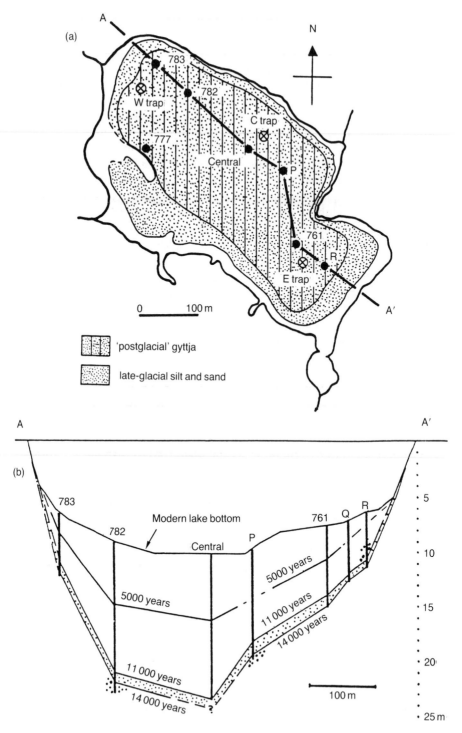

Figure 6.4 (a) Map of Mirror Lake, New Hampshire, showing locations of sediment cores and sediment-trapping sites. (b) Cross-section (along transect A–A') of the basin of Mirror Lake, New Hampshire, showing the temporal pattern of sediment accumulation. From Davis *et al.* (1984).

pattern that reflects the process of 'sediment focusing', that is, the long-term transport of sediment from the littoral zone of the lake's periphery to progressively deeper areas near its geographic center (M. Davis and Ford, 1982; M. Davis *et al.*, 1984). Using five main sediment cores (with radio-carbon dates on the central core, core 782, and core 783 providing an abso-lute chronology for the site; Fig. 6.4) and fifteen supplementary cores (M. Davis *et al.*, 1984), sedimentary, chemical and biological inputs to the whole lake basin have been quantified and expressed as long-term fluxes to Mirror Lake for the time intervals of the late-glacial and the Holocene. Sediment lithology was characterized by particle size and mineralogy, and, corrected for sediment focusing, loss-on-ignition analyses yielded influx rates for deposition of organic matter and mineral grains (Fig. 6.5; M. Davis *et al.*, 1985a, b; Likens and Moeller, 1985a). Chemical extraction and atomic absorption spectrophotometry of the sedimentary sequences documented long-term trends in biochemical compounds (fossilized chlorophyll pig-ments; Likens and Moeller, 1985b) and key chemical elements such as N, P, and S (Fig. 6.5) as well as other major metals found in rock minerals of the watershed (Fig. 6.6; Likens and M. Davis, 1975; Likens and Moeller, 1985a; M. Davis *et al.*, 1985a).

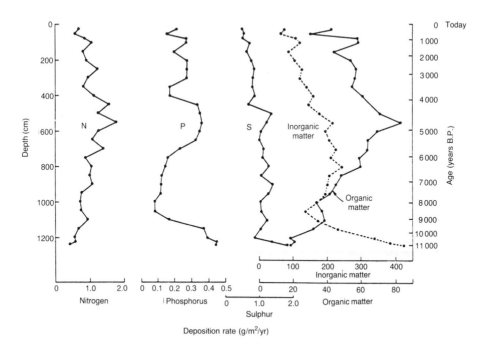

Figure 6.5 Deposition rates of nitrogen, phosphorus, sulfur, inorganic matter and organic matter in the central sediment core from Mirror Lake, New Hampshire. From Likens and Moeller (1985b).

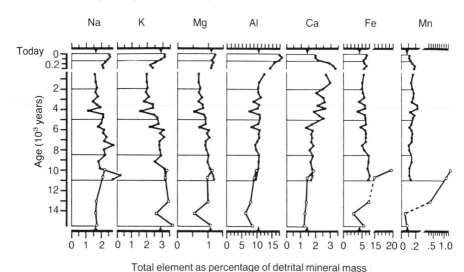

Figure 6.6 Late-glacial and Holocene changes in concentrations of major elements contained in inorganic mineral sediments of Mirror Lake, New Hampshire. Davis *et al.* (1985b).

These paleoenvironmental data support four key observations (M. Davis *et al.*, 1984, 1985a). Influx of mineral grains and rock fragments to Mirror Lake was greatest between 14 000 and 10 000 years ago, accounting for the late-glacial deposition of relatively inorganic, fine sands and silts; in the last 10 000 years, fine-grained organic matter and biogenic opal in diatom frustules have constituted the primary components of the lake mud (gyttja) (Fig. 6.7). Influx of mineral particles, correlated with chemical curves for total Na, K, and Mg (Fig. 6.6), reflects their erosion from uplands of the watershed and transport to the lake by late-glacial winds and geomorphic processes (slopewash and solifluction) and later by streams in the Holocene. In the postglacial history of oligotrophic Mirror Lake, the organic matter preserved in its sediments has largely been produced within the lake basin, rather than transported overland as organic debris from upland slopes. The summary stratigraphic curves for organic matter and the degraded pigments of chlorophyll are interpreted as the aquatic record of long-term changes in lake productivity (Fig. 6.7).

The biological evidence for past terrestrial and aquatic ecosystems consists of fossil assemblages of plant microfossils (pollen grains and spores) and macrofossils of aquatic macrophytes (Likens and Davis, 1975; M. Davis *et al.*, 1985b; Moeller, 1985), diatoms and chrysophytes (Sherman, 1985), animal microfossils of cladocera (Goulden and Vostreys, 1985), and plant pigments (Likens and Moeller, 1985b). Margaret Davis (1985) and colleagues (Likens and Davis, 1975; M. Davis *et al.*, 1985a, b) integrated these

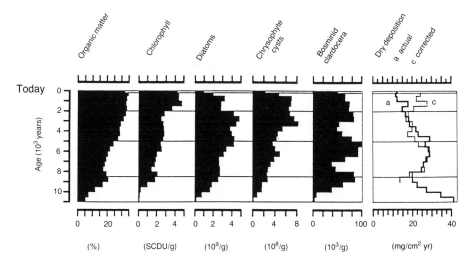

Figure 6.7 Holocene changes in indicators of lake productivity preserved in the sediments of Mirror Lake, New Hampshire. Curves for observed (a) and adjusted deposition rate (c) corrected for sediment focusing, are presented for long-term accumulation of dry mineral matter within the central sediment core. From Davis *et al.* (1985b).

data concerning the paloeolimnology of Mirror Lake and the terrestrial paleoecology of its watershed within the context of six time periods spanning the last 14 000 years.

The late-glacial 'Tundra Period' includes the interval following glacial retreat from the watershed, from 14 000 to 11 500 years ago. The low total influx of pollen (average PAR value of 500 grains $\cdot$ cm^{-2} $\cdot$ yr^{-1} produced primarily by herbaceous vascular plants; M. Davis *et al.*, 1984) indicates a tundra environment with discontinuous patches of sedge, grass, and arctic sage (Fig. 6.8). Herb tundra was replaced by shrub tundra with the invasion of shrub willow, heath, juniper and *Shepherdia canadensis* between 13 000 and 11 500 years ago (M. Davis, 1985). The arctic to subarctic climate, with cold temperatures and high soil moisture, promoted intense frost action, with the alternating freezing and thawing of the glacial till. This periglacial churning of the substrate limited soil development, continually exposed bare ground for invasion by pioneer, shade-intolerant tundra plants, and favored downslope transport of mineral grains by mass wasting. Periglacial solifluction movement of water-saturated sediment, erosion and transport of silt by wind, and stream erosion of fine-grained particles all contributed to the substantial export of mineral grains from the watershed to Mirror Lake. The late-glacial influx of mineral particles into the lake at 13 000 years ago averaged 53 mg $\cdot$ cm^{-2} $\cdot$ yr^{-1}; the late-glacial export value of 650 kg of particulate matter eroded per hectare of drainage area is 32 times

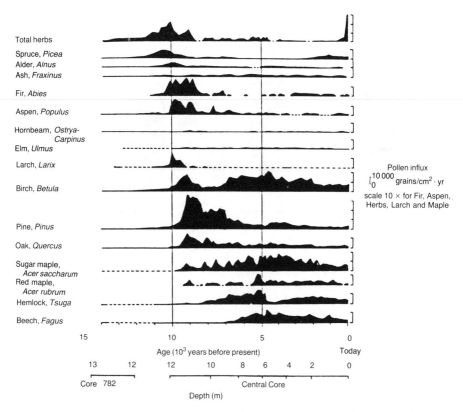

Figure 6.8 Diagram of late-glacial and Holocene changes in pollen accumulation rates for tree genera and total herbs in sediments from Mirror Lake, New Hampshire. From Davis (1985).

greater than the export of particulates from erosion within the modern forested watershed (M. Davis and Ford, 1982; M. Davis *et al.*, 1985a). The continual frost-heaving curtailed development of incipient soils, limited the patchy accumulation of an organic humus layer, and effectively inhibited production of water-soluble nutrients such as phosphorus (P) leached from organic matter. Nitrogen (N) is another key element potentially limiting to plant growth. Late-glacial communities included both *Dryas integrifolia* and *Shepherdia canadensis*, both plants capable of fixing atmospheric nitrogen and contributing this nutrient to soil litter as their phytomass decomposes (M. Davis, 1985).

With the melt-out of ice and formation of the basin 14 000 years ago, Mirror Lake contained clear, alkaline (pH 7.5 to 8.0) water up to 23 meters deep. The very low productivity of aquatic biota in this oligotrophic lake was constrained by the cold climate and severely limiting quantities of available nitrogen. Bottom-dwelling aquatic plants such as marl-encrusting stoneworts (*Chara* and *Nitella flexilis*) served as biogenic sinks for dissolved carbonate leaching from the uplands (Fig. 6.9; M. Davis *et al.*, 1985b; Moeller, 1985).

Between 11 500 and 10 000 years ago, the late-glacial 'Spruce Period' represents the invasion of spruce (Fig. 6.8) into the watershed. This included white spruce (*Picea glauca*), which arrived between 11 500 and 11 300 years ago, as well as the local immigration by 10 000 years ago of red spruce (*Picea rubens*) and, in the nearby White Mountains, black spruce (*Picea mariana*) (fossil evidence of species identifications of spruce is discussed in M. Davis, 1985, and Spear, 1989). Influx values of spruce pollen of 9000 grains $\cdot$ cm^{-2} $\cdot$ yr^{-1} indicate either a late-glacial landscape with stands of spruce established within the tundra mosaic or an open woodland with widely spaced trees. Establishment of spruce trees initiated several ecosystem-level changes (M. Davis, 1985). Subterranean networks of tree roots stabilized hill slopes. Trees altered the hydrologic balance by increasing water loss to the atmosphere through evapotranspiration, lowering the depth to the groundwater table, increasing substrate percolation of precipitation, and reducing the amount of moisture lost overland as run-off of surface water. Fallen spruce logs collected in debris dams across streams, obstructing fluvial transport of sediment and decreasing erosion rates and sediment export (M. Davis *et al.*, 1985b). Periglacial activity, including both massive frost-heaving of substrate and solifluction flow of water-saturated sediment, gradually diminished after 11 500 years ago; the accelerated postglacial warming in temperature reaching modern levels by 10 000 years ago decreased the annual number of freeze-thaw cycles and reduced the effectiveness of cryoturbation (M. Davis *et al.*, 1985b). Between 11 500 and 10 000 years ago, the combination of warming temperatures, dropping water table, and lessened run-off diminished periglacial processes, and debris dams substantially reduced quantities of mineral particles exported from the watershed downslope to the lake. Within the spruce woodland/tundra mosaic, patches of bare ground continued to form along unstable hillslopes until about 10 000 years ago. This shifting patchwork of 'open sites' maintained by geomorphic processes provided opportunities for invasion of new plant migrants on open sites with minimal competition for space and light (M. Davis, 1985). The periglacial disturbance of substrate inhibited the accumulation of continuous humus cover. Incipient soil formation and downward percolation of precipitation increased leaching of soluble carbonate minerals from the soil. Requiring alkaline substrates, calcicolous terrestrial plants such as *Dryas integrifolia* persisted within the watershed until 10 000 years ago, when they were displaced by plants capable of utilizing neutral or acidic soils (M. Davis *et al.*, 1985b). By 10 500 years ago, shrubs of green alder (*Alnus crispa*) established within the spruce woodland. Alder dramatically increased the effectiveness of biological recycling of nutrients in both terrestrial and aquatic ecosystems. Thickets of alder captured atmospheric nitrogen, concentrated it in their phytomass, and their accumulating soil litter was subsequently leached of water-soluble nitrogen compounds that thus leaked from the watershed into Mirror Lake (M. Davis, 1985).

During this late-glacial 'Spruce Period', the gradual transformation from a largely unforested (tundra) landscape to a forested one reflects a shift in the terrestrial ecosystem as both climatic and geomorphic thresholds were crossed (M. Davis, 1985). However, the assemblage of aquatic life in Mirror Lake retained its early character, responding to low nutrient availability, cool temperatures, and water turbidity associated with seasonal influx of mineral silts. By 11 000 years ago, the establishment of aquatic plant populations of quillwort (*Isoetes tuckermani*) on the silty lake mud (Fig. 6.9) and the freshwater diatom flora dominated by a few kinds of plankton (*Fragilaria pinnata*, *Melosira* spp. and *Cyclotella comta*) reflected the alkaline water and low lake productivity (Fig. 6.10; M. Davis *et al.*, 1985b; Sherman, 1985).

Between 10 000 and 5000 years ago during the 'Early Holocene Period', postglacial arrivals of migrants increased the species richness of the mixed

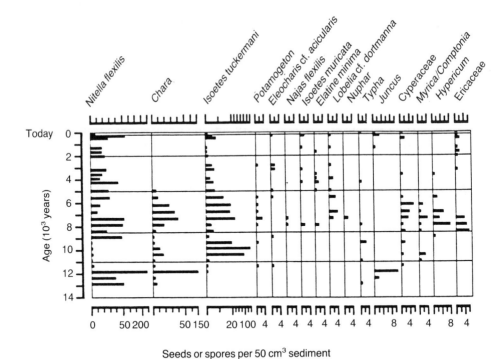

Figure 6.9 Diagram of late-glacial and Holocene changes in plant macrofossil concentrations for seeds, resting spores (*Nitella flexilis* and *Chara*), megaspores (*Isoetes*), and anthers (*Myrica/Comptonia*) of aquatic macrophyte plants preserved in sediments from Mirror Lake, New Hampshire. From Davis *et al.* (1985b).

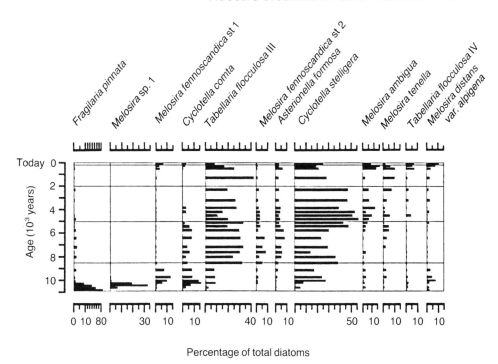

Figure 6.10 Diagram of Holocene changes in diatom assemblages in sediments from Mirror Lake, New Hampshire. From Davis *et al.* (1985b).

coniferous–deciduous forest across the uplands (Fig. 6.8). Following the regional decline in spruce populations after 10 000 years ago, the gaps formed in the forest openings were invaded by shade-intolerant hardwoods such as birch and aspen and other conifers such as balsam fir and juniper (M. Davis, 1985). Local populations of white pine reached peak levels between 9500 and 7000 years ago. By 9000 years ago, oaks, sugar maple and red maple had established in the watershed. With an initial arrival as early as 8500 years ago, hemlock populations achieved peak abundance between 7000 and 5000 years ago. The conversion of birch–aspen–pine forest to northern hardwoods–hemlock forest after 8500 years ago coincided with a series of climatic and environmental changes. Interglacial conditions of warmth were attained by 10 000 years ago. However, peak interglacial temperatures up to 2°C greater than modern values of mean annual temperature were reached between 9000 and 5000 years ago (M. Davis, 1985). Areal expansion of temperate forests after 10 000 years ago substantially increased the volume of water recycled through transpiration and decreased water supply for stream flow. The headward expansion of streams in the increasingly efficient drainage network focused channel incision and concentrated downslope transport of particulate matter along riparian corridors. The combination of fallen logs forming debris dams along streams and the development of a laterally continuous soil–humus layer diminished overall rates of sediment loss from uplands. In Mirror Lake, this is reflected in deep-basin sediments by the changeover in sediment types from late-glacial inorganic silts to early-Holocene organic muds (M. Davis *et al.*, 1985b).

Early-Holocene development of maturing soil profiles increased the available quantities of particulate and dissolved nutrients containing N and P. Accelerated rates of soil leaching between 8000 and 6000 years ago correlates with peak interglacial temperatures and with the replacement of early-successional trees by late-successional, mesic northern hardwoods and hemlock (M. Davis *et al.*, 1985b). Increased nutrient export to Mirror Lake corresponded with the gradual early-Holocene increase in lake productivity, with natural eutrophication shifting from oligotrophic to mesotrophic lake status between 6000 and 5000 years ago. Local extinction of plants along the shallow lake margin (*Chara*; Fig. 6.9) and of small aquatic animals such as cladocera (*Chydorus brevilabris*, typically associated with filamentous algal mats) are interpreted as evidence that a threshold in water quality occurred at about 5200 years ago, with a pH shift from slightly alkaline to slightly acidic water (M. Davis *et al.*, 1985b; Goulden and Vostreys, 1985).

From 5000 to 2000 years ago, during the 'Mid- to Late-Holocene Period', climatic cooling resulted in predominance of northern hardwood forests (Fig. 6.8). Eastern hemlock, which was locally abundant within the watershed 5000 years ago, declined rapidly at about 4800 years ago (M. Davis, 1981b). Within approximately a 30-year span, as represented in Mirror Lake sediments, pollen accumulation rates of eastern hemlock dropped by 90% (M. Davis, 1985). The initial loss of hemlock as a forest-canopy dominant would have increased the area of light gaps in the previously closed forest. The canopy gaps were first captured by birch, which thus replaced hemlock without major physical disruption of the forest soils and hence without large nutrient losses from the terrestrial ecosystem. Over several centuries, birch was competitively displaced as populations of sugar maple, beech, pine and oak expanded. However, by 3000 years ago hemlock populations re-expanded and reached importance values comparable to their earlier levels, 5000 years ago (M. Davis, 1985). Between 5000 and 2000 years ago aquatic communities in Mirror Lake reflected population fluctuations in diatoms (Fig. 6.10), chrysophytes, and bosminid cladocera (Fig. 6.7), but these changes in the aquatic biota appear unrelated to terrestrial dynamics triggered by the decline and recovery of eastern hemlock populations (M. Davis *et al.*, 1985b).

From 2000 until 140 years ago, the 'Late-Holocene Period', mixed conifer–northern hardwood forest occupied the watershed. Late-Holocene increases in spruce (Fig. 6.8) reflect the passage of biological thresholds of temperature tolerance with continued climatic cooling and an increase in annual precipitation to modern levels. The production of acidic litter by spruce promoted the accumulation of soil humus and may have increased the extent of leaching of aluminum from forest soils. After 2000 years ago lake sediments record increased deposition of both aluminum (Fig. 6.6) and chlorophyll concentrations (Fig. 6.7). Enhanced lake productivity and minor acidification of freshwater coincided with diminished populations of diatoms and chrysophytes (Fig. 6.7; M. Davis *et al.*, 1985b; Sherman, 1985).

In the last 140 years of the historic 'Settlement Period' increased pollen values of herbaceous plants such as ragweed (*Ambrosia*), grass and dock (*Rumex*) record regional forest clearance (Figure 6.8). Within the Mirror Lake watershed logging of the forest was followed by a limited extent of farming. Minimal physical disturbance of soil, associated with historic forest clearance (Likens, 1985), produced minor increases in upland erosion and transport of mineral particles to the lake (M. Davis *et al.*, 1985a). The forest soils remained largely intact and sequestered much of the organic reserve of nutrients. With historic regeneration, second-growth forest contains dominants of yellow birch and sugar maple, with reduced populations of spruce, beech, and eastern hemlock. Eastern hemlock was selectively harvested for its tanbark, which was used in the leather-tanning industry (Likens, 1985). Only minor cultural eutrophication of Mirror Lake is indicated by historic expansion of *Melosira* diatoms and increased species richness of diatoms (Sherman, 1985) and the historic decline in overall abundance of cladocera (Goulden and Vostreys, 1985).

The sedimentary record from Mirror Lake is a long-term monitor of ecosystem properties spanning the last 14 000 years, including even the early stages of deglaciation and postglacial landscape evolution. Postglacial invasions of plant populations have established both terrestrial and aquatic communities characterized by long-term increases in species richness and cumulative adjustments in species packing due to environmental changes and competitive interactions. During the late-glacial time interval, from 14 000 to 10 000 years ago, the initial organization of ecosystems (Fig. 6.8) was influenced by three kinds of physical changes on the landscape (M. Davis *et al.*, 1985b): (1) major climatic warming from arctic, through boreal, to temperate conditions; (2) passage of a fundamental geomorphic threshold, with transition from a periglacial environment and massive frost-heaving of incipient soils to a fluvial environment, with progressively stabilized uplands with maturing soils and erosion concentrated within incised channels of streams expanding headward to form the drainage networks of the Hubbard Brook Experimental Watersheds; and (3) stabilization of hillslopes with the invasion of trees into tundra to form the first forests, the development of woody debris jams, and the shift in hydrological balance associated with the transition from tundra to forest (enhanced by the increased evapotranspiration in the forest and by the lowered groundwater table) from overland run-off to increased percolation of precipitation through the soil. The increased soil leaching in turn promoted humus accumulation and maturing of soils and accelerated leaching of dissolved nutrients, such as P and N, the primary terrestrial source for nutrient flux to the aquatic ecosystem of Mirror Lake. The physical presence of trees provided a biological mechanism for reducing the loss of particulate matter from the watershed because of erosion (M. Davis *et al.*, 1985b). On the watershed, closure of the forest canopy at the late-glacial/early-Holocene transition marked the time of shift from an ecosystem organization phase to an aggradation phase (Fig. 6.3).

The beginning of the Holocene interglacial at 10 000 years ago marked the onset of warm climatic conditions and the development of continuous, closed forests across the watershed. M. Davis *et al.* (1985b) interpreted that, from 10 000 to 6000 years ago, early-Holocene forests were 'aggrading' ecosystems (*sensu* Bormann and Likens, 1979; Fig. 6.3), accumulating biomass, establishing a humus layer of organic detritus and progressively acidifying it, and increasing the organic supply of nutrients sequestered in reserve. Enhanced nutrient supply and recycling are linked to the biological fixation of nitrogen from the atmosphere and its cumulative build-up in forest soils, increased production of humic acids associated with decomposition of organic matter, and the accelerated weathering of glacially-derived sediments and the underlying crystalline bedrock. Within these aggrading systems, there are long-term changes in nutrient retention by 'leaky ecosystems'. The early-Holocene early-successional birch–aspen–pine forest was replaced by late-successional northern hardwoods–hemlock forest, and invasions of new plant species to the watershed were completed by about 6000 years ago. With a relatively stable complement of species and with forest ecosystems maintaining relatively high biomass, a prehistoric shifting mosaic of steady-state forest dynamics (Fig. 6.3) was achieved between 6000 and 140 years ago.

In the present interglacial, the cumulative build-up of humus and progressive mineral weathering and maturing of soil profiles have produced no paleoecological evidence for significant, long-term depletion of soil nutrients from the forested watershed of Mirror Lake. Between 10 000 and 6000 years ago, the export of particulate and dissolved nutrients triggered gradual increases in lake productivity; with natural eutrophication during the early Holocene, the aquatic ecosystem of Mirror Lake reached a plateau in productivity ('trophic equilibrium' *sensu* Deevey, 1942) at about 6000 years ago and persisted in this steady state until historic times (M. Davis *et al.*, 1985b).

Thus, in the absence of prehistoric human impact, the early-Holocene aquatic ecosystem of Mirror Lake and the forest ecosystem of its watershed were aggrading systems linked by terrestrial production and export of nutrients to the aquatic sink. In the last 6000 years of the middle and late Holocene, both aquatic and terrestrial ecosystems have been persistent and in steady state, viewed as resilient systems despite species-specific collapse of eastern hemlock populations 4800 years ago. However, on community and population levels, the paleoecological evidence from Mirror Lake records continual fluctuations and readjustments over millennia (M. Davis *et al.*, 1985b).

The model of Iversen (1958), Andersen (1966), and Birks (1986) concerning the cyclic nature of ecosystem development during interglacial periods predicts that during the late-Holocene interval, climatic cooling at temperate latitudes will lead to an 'oligocratic phase' in the absence of human impacts. Iversen's model would propose a corresponding increase in the representation of coniferous trees relative to deciduous trees, accelerated leaching of nutrients from the soil, and subsequent degradation of ecosystems from a mid-Holocene steady state condition. At Mirror Lake, although trends toward both an increase in conifers (spruce) and increased rates of nutrient leaching have been observed, these changes have not yet been of sufficient magnitude to cross a threshold leading to degradation of the Hubbard Brook terrestrial ecosystem (M. Davis *et al.*, 1985b). With continued late-Holocene climatic cooling, however, in the future an oligocratic phase would be expected to ensue (Birks, 1986). In this eventuality, the shifting-mosaic steady state that developed in the last 6000 years on the Hubbard Brook watersheds would no longer exist.

6.4 NATURAL AND ANTHROPOGENIC ECOSYSTEM CHANGES IN SWITZERLAND

The case study of Lobsigensee, Switzerland, illustrates the combined use of many paleoecological techniques in reconstructing ecosystem-level changes on a coupled watershed-lake system. In a series of papers (Ammann *et al.*, 1983; Ammann, 1986, 1988, 1989a, b); Ammann and Tobolski, 1983; Elias and Wilkinson, 1983; Hofmann, 1983; Chaix, 1983; Gaillard, 1983), Brigitta Ammann and her colleagues have used pollen assemblages, plant macrofossils, beetle, chironomid and molluscan faunas, and analysis of sediment texture and chemistry to decipher the history of ecosystems surrounding the Lobsigensee lake site since the last glacial retreat. Lobsigensee is a small (2 hectares, maximum depth 2.5 meters), closed lake basin located on the till-covered Swiss Plateau. The site was covered by ice of the Rhone glacier during the last, or Weichselian, glaciation. The vegetation in the lake today is composed of floating-leaved aquatics (*Nymphaea*) and reeds (*Phragmites*), with a narrow belt of riparian alder forest (*Alnus glutinosa*) surrounding the lake and with cultivated fields, extending beyond. Before agriculture, the natural vegetation of the surrounding uplands was beech (*Fagus sylvatica*) forest (Ammann and Tobolski, 1983).

Based upon a series of sediment cores taken on a transect across the lake basin, it was determined that during the late-glacial interval, Lobsigensee was at least 10 hectares in surface area and at least 17 meters deep. Earliest lake sediments, dating from more than 14 000 years ago, contain fossil evidence of a periglacial environment with only sparsely colonized, open ground, as indicated by low pollen accumulation rates (PAR) of primarily *Artemisia* and the absence of algae and aquatic vascular plants. By about 14 000 years ago, a pioneer flora began to establish on the till plain surrounding the site, with heliophilous herbs predominating; aquatic plants included pondweeds (*Potamogeton*) and water milfoil (*Myriophyllum*) as well as algae such as *Pediastrum* (Ammann and Tobolski, 1983). The fossil insect fauna (Elias and Wilkinson, 1983) included both terrestrial and aquatic species today distributed in open-ground environments of arctic tundra in northern Fennoscandia and high elevations of the Swiss Alps. Littoral zone chironomids also reflected in composition the cold periglacial environments of this 'Oldest Dryas' interval (Hofmann, 1983). The snail fauna from this time interval was relatively poor in number of species (10), and dominated largely by the genus *Pisidium*, reflecting the harshness of the cold, continental climate (Chaix, 1983).

After about 13 500 years ago arctic-alpine shrubs, including dwarf birch (*Betula nana*), established near the shores of Lobsigensee, as documented both by pollen (Gaillard, 1983) and plant-macrofossil evidence (Ammann and Tobolski, 1983) from the lake sediments. Both alpine herbs (e.g. *Saxifraga oppositifolia*, *Plantago montana*, *Plantago alpina* and *Oxyria*) and steppe species (*Ephedra*) continued to occupy open-ground environments. With increasing vegetative cover on the watershed and with aquatic vascular plants occupying the lake, erosion of silicate minerals declined, and the increasing productivity of the lake resulted in deposition of more carbonate-rich sediments in the basin (Ammann and Tobolski, 1983). Local presence of *Salix* and *Betula* shrubs around the lake margins was indicated by presence of fossil leaf beetles (Elias and Wilkinson, 1983).

After about 13 000 years ago, marked changes occurred in both terrestrial and aquatic ecosystems at Lobsigensee. PAR values increased rapidly, the composition of pollen spectra changed from predominantly herbs to shrubs and trees, including tree birch (*Betula alba*). Heliophilous pioneer species of both *Juniperus* and *Hippophae rhamnoides* became established on the watershed of Lobsigensee (Ammann and Tobolski, 1983). *Hippophae* is a shrub with symbiotic actinomycetes in its root nodules that enable it to fix nitrogen and hence colonize nutrient-poor soils. The cold-requiring species of insects were replaced by temperate species indicative of both increasing summer warmth and establishment of swamp vegetation around an increasingly eutrophic lake (Elias and Wilkinson, 1983). A late-glacial combination of warming temperatures and decreasing water depth resulted in a change from strictly aquatic chironomid faunas to semi-aquatic forms (Hofmann, 1983). Sediments changed from clay-rich to carbonates, favoring an increase in numbers of individuals of snails that could be supported in the lake (Chaix, 1983).

Pine (*Pinus sylvestris*) became the predominant tree in a forested landscape around Lobsigensee after 12 000 years ago. After 10 000 years ago, deciduous trees immigrated, and they became dominant after 9000 years ago (Ammann and Tobolski, 1983). This early-Holocene transition to temperate conditions after 10 000 years ago was also reflected at Lobsigensee by a change in lake sediments from carbonate-rich marl to peat (Ammann and Tobolski, 1983).

Although in general the major changes in terrestrial and aquatic ecosystems at Lobsigensee appear to have followed closely the nearly stepwise changes in regional climate through the late-glacial and early Holocene intervals, Ammann (1989a, b) notes that differential response times of species to climatic changes resulted in lags in establishment of certain species. For instance, temperate beetles and aquatic macrophytes such as cattail (*Typha latifolia*) established at Lobsigensee with little time lag following first climatic amelioration. The beetles are not dependent upon the prior establishment of particular species of plants and have a fast reproductive rate that allows them to disperse rapidly. Similarly, a combination of annual generation time, efficient seed dispersal by waterfowl, and independence of soil development account for the rapid spread of *Typha* into aquatic sites on deglaciated terrain. However, temperate shrubs such as hazel (*Corylus*) lagged as much as 3500 years behind beetles with similar modern-day distributions in establishing their local populations around Lobsigensee. Ammann (1989a, b) interprets this apparent time lag as the result of slow migration of hazel populations. Hazel may have been limited in its rate of migration because of poor dispersal of its heavy fruits, as well as life-history characteristics such as a relatively long time required between seed germination and successful flowering of the mature plant (a minimum of 10 years). An additional factor in the delayed migration of *Corylus* may have been its sensitivity to late-winter frosts that may not have been a constraint on other species (Ammann, 1989a, b).

At Lobsigensee, the onset of peat formation occurred first in the littoral zone (during the transition from late-glacial to Holocene conditions), and later toward the center of the basin as the lake filled in with sediment and lake levels fell (Ammann, 1986). In the lake sediments, laminations consisting of alternating organic layers (representing summer aquatic productivity) and carbonate layers (deposited in winter) were found in two zones, one layered sequence corresponding with the late-glacial interval and the other occurring in the mid-Holocene. In the first case, the laminations are explained by warming temperatures, falling and fluctuating lake levels, and overgrowth of the shallow lake margin resulting in smaller lake size, natural eutrophication of the lake, and subsequent reduction in lake circulation that kept the seasonally deposited sediment types from being reworked. The second interval of lake eutrophication and deposition of laminated sediments corresponded with settlement of Neolithic farmers in Switzerland about 5000 years ago. Local Neolithic settlement and agriculture on the shores of Lobsigensee was reflected in the fossil record by charred seeds of wheat (*Triticum aestivum*). In this case, formation of layered lake sediments was probably induced by a chemical reaction that changes the bicarbonate equilibrium in the water. An increase in phosphorus concentration in Lobsigensee because of human occupation of its watershed would produce carbonate sediments, as phytoplankton consumed the phosphorus and produced a supersaturated solution out of which calcite crystals would have precipitated in the winter season. The paleolimnological record from Lobsigensee thus demonstrates two intervals of eutrophication, one natural and one anthropogenic, that changed the trophic status of the lake markedly (Ammann, 1986; 1989b).

6.5   MAYAN IMPACT ON WATERSHEDS IN THE PETEN

The role of human activities in affecting the movement of materials from upland watershed to lake has been studied through paleoecological investigations in the lowlands of the Peten region of Guatemala (Binford *et al.*, 1987). These paleolimnological investigations were conducted on coupled lake-drainage basin ecosystems located in a region of Central America today characterized by tropical lowland dry forest. The primary objective was to evaluate ecosystem-scale changes in fluxes of material from the watershed to the lake sediments in relation to cultural disturbances correlated with prehistoric settlement of the Peten region.

Earliest Mayan settlement in the study area dates from about 1000 B.C. (about 3000 years ago), as represented by house mounds of the Middle Preclassic period. Changes in human population density were estimated for each archaeological period by mapping of house mounds. With the assumption that 84% of the mapped mounds were residential structures and that 5.6 people lived in each house, the resulting estimates of population density for the culmination of each cultural period yield consistent trends for five watersheds examined in the Peten region (Fig. 6.11). In general, human populations increased from the Middle Preclassic (1000–250 B.C.) through the Terminal Preclassic (100 B.C. to A.D. 250) to the Early Classic (A.D. 250 to 550), although on some watersheds populations declined during Terminal Preclassic times. By the end of the Late Classic period (A.D. 850), however, the region was entirely reoccupied by large human populations, between 3 and 14 million individuals.

The Mayan civilization collapsed after A.D. 850, either because of human interactions such as warfare or spread of disease, or because of environmental changes due to natural causes, possibly including earthquakes or soil degradation from intensive agriculture (Binford *et al.*, 1987). Evidence for deterioration of agricultural soils is found in the sediment geochemistry of soil developed on culturally terraced hillslopes (Healy *et al.*, 1983). Soils developed within Mayan agricultural terraces dating from Early to Late Classic times yield pollen of maize (*Zea mays*) as well as high percentages of

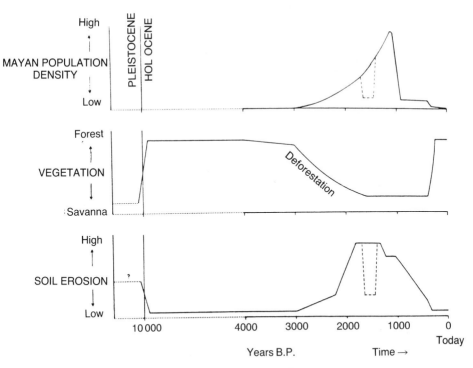

Figure 6.11 Summary diagram of long-term impacts of Mayan settlement on the vegetation and soils of terrestrial environments in the Peten lowlands of Guatemala. Dashed curves represent fluctuations of human populations on several watersheds and their short-term influence upon soil erosion. From Binford *et al.* (1987); reprinted with permission from Pergamon Press.

clay in the sediment and high elemental concentrations of aluminum and manganese. Although upper slopes of terraced hills were apparently left vegetated and not used for agriculture, forest clearance and terracing of lower slopes resulted in accelerated run-off and colluviation of sediment that eventually filled in the low areas behind stone terrace walls, accounting for accumulation of both clay-sized particles and toxic elements through time. Based on archaeological evidence for high, Late Classic population densities near the terraced zones, Healy *et al.* (1983) suggest that human development of hillslope terraces was essential for supporting intensive maize cultivation. Long-term intensive cultivation on the terraces, combined with hillslope erosion, resulted in changes in physical and chemical characteristics of the soil that eventually led to declining yields of agricultural crops over time. After the collapse of the prehistoric Mayan civilization, population densities in the Peten remained low until the present century.

Paleoecological analysis of sediments from Peten lakes shows that late-glacial vegetation of the region was a mosaic of marsh, savanna, and juniper scrub (Leyden, 1984). Temperate forest replaced savanna and scrub after 10 750 years ago, then was rapidly replaced by mesic tropical forest that persisted until about 3000 years ago (Fig. 6.11). Coincident with first Mayan occupation, forest trees declined in representation in the fossil record relative to taxa characteristic of grasslands and human-disturbed areas (e.g. Ambrosiae). High forest regenerated after collapse of the Mayan civilization (Fig. 6.11), as indicated by a shift in pollen assemblages from dominance of non-forest to forest taxa (Binford *et al.*, 1987).

A change in lithological composition of lake sediments occurred at the time of first forest clearance in the Middle Preclassic period, about 1000 B.C. In these lakes, net accumulation rates increased for both phosphorus and for

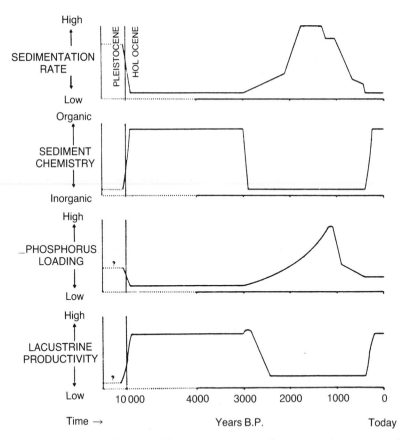

Figure 6.12 Summary diagram of long-term impacts of Mayan settlement on sedimentation rates, sediment chemistry, phosphorus concentrations and productivity of lakes in the Peten region of Guatemala. Modified from Binford *et al.* (1987); reprinted with permission from Pergamon Press.

mineral particles of siliceous silts and clays and of carbonates (Fig. 6.12). After Mayan abandonment of the region, organic-rich sediments once again began to accumulate (Binford *et al.*, 1987). During the interval of occupation by Mayan people, rates of inputs of phosphorus to the lakes of the Peten were of the order of magnitude expected from human physiological output (0.5 kilograms of phosphorus per person per year; Fig. 6.12). This accelerated transport of phosphorus to the lake was probably accomplished by transport of sediments within streams or reworked across hillslopes from the surrounding watershed. Contrary to expectations based on studies of fertilization of modern lakes by phosphorus- and nitrogen-rich sediments from agricultural land, Mayan forest clearance and agriculture did not cause the lakes to become eutrophic. Rather, the high prehistoric input of inorganic sediments that accompanied increased phosphorus loading rapidly buried the nutrients in the sediment 'sink', making them unavailable for use by aquatic organisms. In addition, high turbidity of the water caused by accelerated erosion of fine-grained mineral particles may have reduced aquatic productivity, accounting for the apparently lowered rates of organic matter accumulation relative to both pre- and post-Mayan occupation times (Fig. 6.12; Binford *et al.*, 1987).

Paleoecological studies in the Peten region demonstrate that prehistoric human activities have been significant in affecting ecosystem-level fluxes of nutrients from terrestrial to aquatic environments. In the case of the Maya, deforestation, intensive agriculture, and activities related to construction of cities increased the rates of soil erosion from uplands and its transport downslope into the man-made agricultural terraces and natural lakes. Nutrients were delivered in greatly increased silt loads to the lake basins, but in a form unavailable to aquatic plants; hence the nutrients were removed from the coupled watershed-lake system and buried in a long-term nutrient sink within the lake sediments. Mayan agriculture on terraced hillslopes resulted in long-term soil degradation and accumulation of toxic elements. Prehistoric human activities on an overpopulated landscape thus may have led to deterioration of both the local terrestrial and aquatic environment upon which the Maya depended for continued survival. Negative ecological feedbacks on an ecosystem level thus may have been an important factor contributing to the eventual decline of the Mayan civilization. The destabilizing effect of increasing prehistoric human land use on ecosystem dynamics is an example of threshold effects that have repercussions at all trophic levels.

6.6 PALEOLIMNOLOGICAL RECORD OF LAKE EUTROPHICATION, PALUDIFICATION AND ACIDIFICATION

Three processes characterize the long-term development of many lake ecosystems within the northern temperate zone (Deevey, 1984; M. Davis

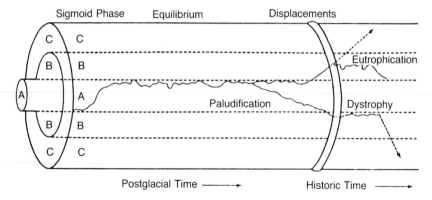

Figure 6.13 Hypothetical model of the stability of trophic equilibria in space and time based on the interglacial development of eutrophic lakes such as Linsley Pond, Connecticut, dystrophic lakes such as Cedar Bog Lake, Minnesota, and culturally eutrophied lakes such as Lake Washington, Washington State. From Deevey (1984).

et al., 1985a, b; R. Davis, 1987): (1) eutrophication; (2) paludification; and (3) acidification. Trajectories of lake ecosystem development may be driven by nutrient loading, that is, the flow of material through aquatic ecosystems. Changes in nutrient loading drive overall changes in lake productivity and trophic level interactions. The balance of nutrient loading is tied to (1) nutrient availability, (2) the residence time for recycling of nutrients within the aquatic ecosystem, and (3) nutrient losses from the system, for example, as they are buried in inaccessible forms within the sediment sink or flushed from the lake by stream outflow (Deevey, 1984).

Postglacial eutrophication results from a long-term increase in nutrient loading, with the aggradation of aquatic biomass and increased build-up and recycling of nutrients (Fig. 6.13). For example, for Linsley Pond, Connecticut, the postglacial lake system shifted from nutrient-poor, oligotrophic status to nutrient-rich, eutrophic status, as indicated by the early-Holocene rise to dominance of *Melosira* populations in the diatom flora about 8000 years ago (Patrick, 1943; Brugam, 1978b). Linsley Pond achieved trophic equilibrium for much of the Holocene; human disturbance including deforestation and local farming triggered the accelerated flux of nutrients to the pond and fostered a pulse of cultural hyper-eutrophication in the last three hundred years (Brugam, 1978a, b).

Long-term shifts in nutrient loading, for example, a reduction in nutrient availability, may result in a dystrophic lake ecosystem (Fig. 6.13). In this case, paludification of the lake may result from sediments filling in the basin and from cumulative sequestering of nutrients within the biomass of accumulating peat, as demonstrated at Cedar Bog Lake, Minnesota (Lindeman, 1941, 1942). Over the last glacial–interglacial cycle, there may be several changes in trophic state in the long-term development of a freshwater lake. More than one period of eutrophication may have been caused by both natural and anthropogenic processes, as demonstrated at Lobsigensee, Switzerland (Ammann, 1986; 1989b). In the paleolimnological record, changes in trophic status of lakes are typically inferred from sedimentary changes in the accumulation of organic carbon. Whiteside (1983) cautions that prehistoric changes in influx rates for inorganic sediment may produce the observed trends, and that many forms of physical, geochemical, and biotic evidence are required in order to separate the effects of changes in trophic status of a lake from those of other causes when interpreting the paleolimnological record. For example, in the case of the influence of Mayan people on coupled watersheds and lakes in the Peten (Binford *et al.*, 1987), anthropogenic activities resulted in accelerated erosion of the uplands and in the deposition of substantial quantities of mineral particles into the lakes, either depressing aquatic productivity through increased turbidity or masking effects of increased productivity as would be observed in the sedimentary curve for organic carbon accumulation (Fig. 6.12).

Fossil diatom assemblages (class Bacillariophyceae) provide evidence for long-term changes in aquatic productivity, paludification, and lake chemistry including acidification caused by both natural and cultural processes (Brugam, 1983; R. Davis, 1987; Ford, 1990). Inferences of long-term changes in pH of lake water can be quantified on the basis of pH preferences or tolerances of modern diatom species and diatom groups and applying this knowledge to fossil diatom assemblages. Diatom-based reconstructions of lake acidity are based upon calibrations of extant diatoms and pH measurements of the aquatic habitats in which they live. As used in paleolimnological studies, such calibration data sets typically include an array of modern lakes that exhibit the full range of water chemistry conditions likely to be represented in the prehistoric sequence of lake development, as reflected in the fossil diatom assemblages preserved 'downcore' in lake sediments (R. Davis, 1987). Recent concerns about acid deposition and its potential impacts upon both terrestrial and aquatic ecosystems make paleolimnological studies of long-term trends in lake acidification essential because of the lack of direct measurements of lake pH prior to A.D. 1920 (R. Davis, 1987).

Dick Brugam (1980, 1983) used cluster analysis to compare the composition of modern assemblages of diatoms collected from 105 lakes in Minnesota with the fossil record of diatoms from Kirchner Marsh, east-central Minnesota. Based upon this quantitative comparison of modern and fossil diatoms from the western Great Lakes region, Brugam (1980, 1983) reconstructed changes in lake chemistry, including total alkalinity and water transparency, as well as water depth at Kirchner Marsh through the postglacial interval. Brugam (1980, 1983) found that diatom samples from the mid-Holocene interval at Kirchner Marsh were most similar to those collected today from shallow lakes located in the prairie region of southwestern Minnesota. From this observation, he concluded that the Kirchner Marsh site experienced a time of shallowing water depth during the warmer and drier climatic conditions of the mid-Holocene interval (Brugam, 1980, 1983). Late-Holocene increases in *Melosira distans* were interpreted as a biological response to acid, boggy conditions in the marsh, indicating both acidification of the lake water and gradual filling in of the kettle-lake basin with sediments (Brugam, 1980, 1983).

Charles (1985) compared several kinds of diatom-pH calibrations as applied to lakes situated in the Adirondack Mountains of New York. Calibration techniques were based on (1) pH preference of diagnostic or indicator diatom species; (2) inferences based upon pH range limits of five diatom groups proposed by Hustedt (1939); and (3) a variety of numerical indices derived from the biological data (Charles, 1985). For example, in the high peaks region of northern New York State, euplanktonic species of diatoms such as *Melosira ambigua* and *Cyclotella comta* are nearly exclusively found in lakes with a limiting threshold of water pH at or above 5.8. These calibrations were applied to three lakes situated on an altitudinal transect in the Adirondack Mountains (Whitehead *et al.*, 1986). These lakes included Heart Lake, situated at 661 meters elevation and surrounded by mixed conifer–northern hardwoods forest; Upper Wallface Pond, located at 948 meters elevation, within subalpine forests of red spruce (*Picea rubens*), balsam fir (*Abies balsamea*) and paper birch (*Betula papyrifera*); and Lake Arnold, at 1150 meters elevation, within a watershed forested by red spruce and balsam fir. Several different kinds of modern calibrations were applied to all sites, and, based on average values of diatom-inferred pH, curves for the pH of lake water were plotted for the past 12 500 years (Fig. 6.14; Whitehead *et al.*, 1986). Following the retreat of glacial ice and the formation of the lake basins at about 12 500 years ago, changes in diatom assemblages reflected a major decline in the pH of the lake water at all sites. By 8000 years ago, pH stabilized at approximately 5.0 at Lake Arnold and at 5.5 at Upper Wallface Pond. Heart Lake acidified progressively, reaching pH values of 5.9 by 6000 years ago. The maximum rate of pH decline (0.2 to 0.4 pH units per 100 years) occurred about 10 000 years ago during the transition from late-glacial to Holocene conditions. This rapid acidification of lake water was associated with postglacial climatic warming, increased rates of weathering of fresh till, leaching of the incipient soil, and water transport of base cations to the lakes. With the change in upland vegetation from tundra to spruce woodland and with the development of humus-rich soils, terrestrial plant biomass increased on all three watersheds and contributed to long-term acidification with a net uptake and accumulation of cations (Whitehead *et al.*, 1986). The average Holocene pH in lake sites decreased with increasing elevation of the sites, reflecting altitudinal gradients in watershed biogeochemistry influenced by gradients in dominance of conifers, thickness of soil mantle, rates of weathering, and precipitation availability (Whitehead

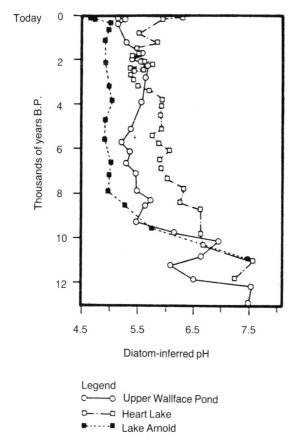

Today

Thousands of years B.P.

Diatom-inferred pH

Legend
○——○ Upper Wallface Pond
□—·—□ Heart Lake
■·····■ Lake Arnold

Figure 6.14 Late-glacial and Holocene changes in freshwater pH inferred from diatom assemblages in sediments of three lakes in the Adirondack region of New York State. From Whitehead *et al.* (1986).

et al., 1986). In sediments of historic age, the population increase in *Fragilaria acidobiontica* and the decline in diatom-inferred pH indicate further lake acidification in Lake Arnold and Upper Wallface Pond that can be attributed to acid deposition from the atmosphere (Whitehead *et al.*, 1986).

At a fourth lake site within the Adirondack Mountains, Big Moose Lake (Fig. 6.15), historic sediments dated by Pb-210 provide a record of accelerated acidification (R. Davis, 1987; Charles *et al.*, 1987). Diatom assemblages preserved in historic sediments indicate that lake waters were acidic, with pH of 5.5 to 6.0, from the A.D. 1800s to the mid-1950s. However, a major pulse of lake acidification between the mid-1950s and the 1980s resulted in a drop in pH to 4.9. This decrease in pH was associated with a loss of a large number of planktonic diatom species. This recent pulse of

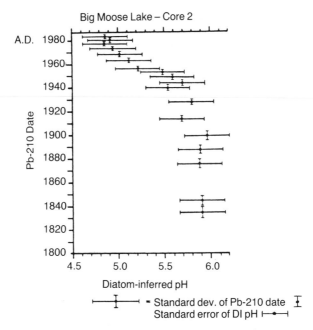

Figure 6.15 part text...

Figure 6.15 Historic changes in freshwater pH inferred from diatom assemblages in sediments of Big Moose Lake, New York. From R. Davis (1987); reprinted with permission from D. F. Charles and Pergamon Press.

acidification was preceded in the first half of the 20th century by sedimentary evidence for a rise to high concentrations of chemical elements including Cu, Pb, and Zn and of particulate matter including soot and polycyclic aromatic hydrocarbons. These biological, sedimentary and geochemical forms of evidence are proxy indicators of historic acid emissions generated during burning of fossil fuels and deposited in the lake basins by acid rain (R. Davis, 1987; Charles *et al.*, 1987). Charles *et al.* (1987) conclude that the historic lake acidification inferred from diatom assemblages is the result of regional acid deposition from the atmosphere and cannot be attributed to events of disturbance within the watershed.

Marjorie Winkler (1988) studied the late Pleistocene and Holocene history of Duck Pond, located on Cape Cod, Massachusetts. This oligotrophic lake is situated in an area underlain by nutrient-poor, quartz-rich sandy outwash. Postglacial vegetation has been dominated by a succession of conifers, including spruce (*Picea*), jack pine (*Pinus banksiana*), white pine (*Pinus strobus*) and pitch pine (*Pinus rigida*). At Duck Pond, the diatom-reconstructed pH has averaged 5.3 ± 0.3 throughout the last 12 000 years, indicating stability of paleolimnological conditions within a long-term, naturally acidic watershed and lake ecosystem. Winkler (1988) suggested that changes in forest composition, forest fire regime, and in historic times the deposition of wind-borne industrial acids have resulted in subtle fluctuations in the pH of lake water. Within the last 150 years, the combination of increased forest fires and acid rain have produced an increased acidity in Duck Pond, reaching a pH of 5.1 ± 0.1 (Winkler, 1988). Thus the vulnerability of a lake ecosystem to acidification, either naturally or culturally induced, is inherently tied to the starting conditions of the substrate within the surrounding watershed and its buffering capacity.

6.7 EXTINCT ECOSYSTEMS – THE PLEISTOCENE ARCTIC STEPPE–TUNDRA HYPOTHESIS

Based upon a number of studies of fossil pollen and plant-macrofossil assemblages and vertebrate faunas across the region of Beringia, extending from eastern Siberia to Alaska and northwestern Canada (references cited in Hopkins *et al.*, 1982), the hypothesis has been proposed that a major late Pleistocene ecosystem, the Arctic steppe–tundra, is now extinct. This former ecosystem is defined by a tundra vegetation dominated by species of grass (including the gramma grass, *Bouteloua*, today typical of short grass prairie), sedges, and sage (*Artemisia*) that supported a diverse herbivore fauna primarily composed of ungulates (Bliss and Richards, 1982; Guthrie, 1982, 1984). Late-Pleistocene arctic paleoenvironments were characterized by a drier and colder climate than today, but with warmer summers that allowed deeper seasonal thawing of the permafrost (Bliss and Richards, 1982). Wind-blown deposits of silt (loess) provided a fresh, unweathered source of nutrients that were continually replenished to the soil and that were made available to vegetation during the growing season because of the thawing permafrost that deepened the active layer of the soil (Guthrie, 1982, 1984).

Great differences occurred in composition and functional groups of arctic vertebrates between the late Pleistocene and today (Guthrie, 1984; Hopkins *et al.*, 1982). For example, the Pleistocene arctic fauna of Alaska contains up to 95% grazing animals such as bison (*Bison*), horse (*Equus*) and mammoth (*Mammuthus*) in a region today characterized by tundra vegetation (Guthrie, 1984). The fauna included taxa now found in temperate grasslands such as badgers (*Taxidea*) and ferrets (*Mustela*), along with Asiatic saiga antelope (*Saiga*) and lion (*Panthera*), North American short-faced bears (*Arctodus*) and camels (*Camelops*), as well as sabertooth cats (*Homotherium*), yaks (*Bos*), and bonnet-horned musk oxen (*Symbos*) (Guthrie, 1968, 1984). Large mammals found in the fossil assemblage that still coexist in Alaska include moose (*Alces*), caribou (*Rangifer*), musk oxen (*Ovibos*) and sheep (*Ovis*) (Guthrie, 1968, 1984). The high species richness of the ungulate animal community of the late Pleistocene steppe–tundra has been explained as the result of a more fine-grained and complex mosaic of plant communities than today. Greater

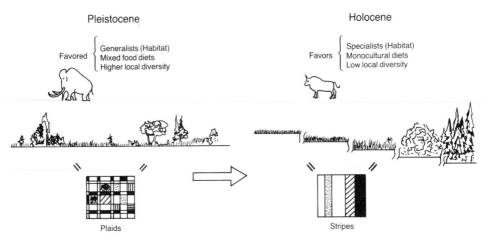

Figure 6.16 Schematic representation of changes in the Alaskan landscape mosaic from Pleistocene to Holocene time and their effects on populations of large herbivores. (a) Pleistocene climates favored a higher local diversity of plant species which favored large herbivore species that used mixed food diets and made use of a broad range of habitats. (b) Less varied growing seasons of the Holocene resulted in a greater biotic zonation, favoring simplified faunal assemblages that are more precisely adapted to a specific habitat and that can survive on a less varied diet. From Guthrie (1984).

diversity of plant communities, and higher net primary productivity made possible by longer, warmer growing seasons and increased nutrient availability, together with a selection of grassland (especially graminoid) species that were more palatable than the tussock sedges such as cotton grass (*Eriophorum*) that predominate over extensive areas today, would have reduced competition among large herbivores and allowed them to coexist (Fig. 6.16; Guthrie, 1982, 1984). Dale Guthrie (1982, 1984) has proposed that major changes in plant community organization, habitat diversity, landscape heterogeneity, and trophic level interactions at the end of the Pleistocene were important contributing factors to the widespread extinctions of many of the herbivore genera.

Using pollen evidence for changes in relative abundances of important plant taxa, Bliss and Richards (1982) reconstructed the proportion of arctic landscape in each of four plant communities (Fig. 6.17). Habitat types included (1) floodplains and lower river terraces with tall shrub communities dominated by species of willow (*Salix*), with an understory rich in herbs and grasses in openings (habitat representing 5% of the total area); (2) poorly drained lowlands supporting sedge-moss meadows (10% of the total area); (3) well-drained, nutrient-rich, rolling uplands covered with a mosaic of upland sedges, grasses, and *Artemisia* (65% of the total area); and (4) wind-exposed, relatively dry mountain slopes supporting tundra cushion plants

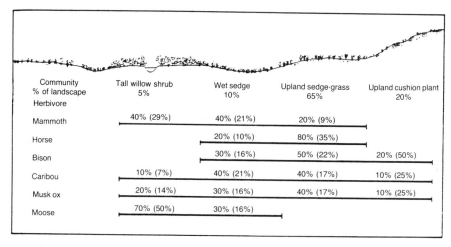

Community % of landscape Herbivore	Tall willow shrub 5%	Wet sedge 10%	Upland sedge-grass 65%	Upland cushion plant 20%
Mammoth	40% (29%)	40% (21%)	20% (9%)	
Horse		20% (10%)	80% (35%)	
Bison		30% (16%)	50% (22%)	20% (50%)
Caribou	10% (7%)	40% (21%)	40% (17%)	10% (25%)
Musk ox	20% (14%)	30% (16%)	40% (17%)	10% (25%)
Moose	70% (50%)	30% (16%)		

Figure 6.17 Reconstruction of the 25 000 yr B.P. Arctic-steppe Mammoth ecosystem of Beringia (western Alaska and eastern Siberia) showing dominant plant communities and six dominant herbivore species. The first value represents the percentage of the total forage that a given herbivore obtained from a given plant community (add across the row for all communities to account for the animal's diet). The second figure (in parentheses) is the percentage contribution that a given community provided to the total diet of major herbivore species (add down the column to characterize the herbivore assemblage utilizing the same plant community). From Bliss and Richards (1982).

including *Dryas, Saxifraga,* and *Draba,* along with dry-site sedges and lichens (20% of the total area) (Bliss and Richards, 1982). For six important herbivore species, estimates were made for both the percent of total forage obtained from each plant community and the percentage contribution of each plant community to the total diet of herbivore species (Fig. 6.17; Bliss and Richards, 1982). Based on a landscape unit of 1000 km^2 and using data on net primary production from modern arctic ecosystems, Bliss and Richards (1982) reconstructed the theoretical food web of the late Pleistocene steppe–tundra ecosystem (Fig. 6.18). Energy flow relationships thus determined demonstrated the important role not only of the upland sedge–grass–*Artemisia* community but also of the landscape diversity afforded by the mosaic of tall-shrub and wet sedgeland habitats (Bliss and Richards, 1982). In the now-extinct ecosystem, calculated efficiency of biomass transfer (standing crop divided by net production consumed) from vegetation to herbivores and from herbivores to carnivores (humans) was 0.4% and 0.1%, respectively, each value a full tenth of 1% greater than that calculated for the modern tundra ecosystem on the high Arctic Truelove Lowland (Bliss and Richards, 1982).

Several of the premises upon which reconstructions of the extinct late

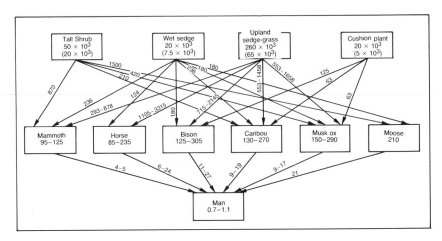

Figure 6.18 Diagram of energy flow of the 25 000 yr B.P. Arctic-steppe Mammoth ecosystem of Beringia (western Alaska and eastern Siberia). Values in the boxes are live above-ground biomass and net annual production (values in parentheses) for the plant communities and the live biomass of herbivores and people. Transfers between boxes have values representing net production consumed per year. Units are metric tons per 1000 km². From Bliss and Richards (1982).

Pleistocene steppe–tundra ecosystem were made have been questioned (Cwynar and Ritchie, 1980; Ritchie and Cwynar, 1982). Evidence from studies of fossil pollen accumulation rates in several arctic lakes indicates that during the late Pleistocene, productivity of pollen was much lower than at present, from which it can be inferred that the apparently high representation of arctic sedges, grasses and sage (*Artemisia*) in many late Pleistocene pollen diagrams may be an artifact of percentage calculations rather than a true increase in vegetative cover or in net primary productivity of these taxa (Cwynar and Ritchie, 1980; Ritchie and Cwynar, 1982). Further, many of the taxa found in trace amounts in pollen diagrams are tundra cushion plants or other taxa indicative of fell-field vegetation. Rather than a complex vegetation mosaic in which upland sedge–grass–sage communities were important, Cwynar and Ritchie (1980) and Ritchie and Cwynar (1982) proposed that the late Pleistocene landscape of Arctic Northwestern Canada was a sparse tundra or polar desert. As an alternative to this hypothesis, Guthrie (1982) suggested that the apparently low pollen accumulation rates of sedges and grasses may have resulted from suppression of flowering and stimulation of vegetative growth by intensive grazing activities of the rich assemblage of herbivores. Resolution of the controversy will require additional research with integrated paleoecological studies that include plant-macrofossil analysis for refined taxonomic resolution of sedge and grass species (Matthews, 1982) and that are conducted at the scale of the hypothesized former landscape mosaic for appropriate identification of plant communities.

During the late Pleistocene, the unglaciated portion of the high Arctic region was characterized by a mosaic of plant communities that was sufficiently complex and productive that it apparently supported a highly diverse herbivore fauna with maximum species packing and minimal competitive exclusion. As reconstructed by the majority of arctic paleoecologists (Hopkins *et al.*, 1982), this ecosystem differed from any observable today. If the hypothesized extinct Pleistocene Arctic steppe–tundra ecosystem is verified, it adds new insights into generalizations concerning both the structure and functioning of ecosystems. This example demonstrates that the field of Quaternary ecology can contribute important empirical evidence with which to test the properties of ecosystem stability and resilience predicted by food-web models (Pimm, 1982). The structure of arctic plant communities and landscape heterogeneity during the late Pleistocene directly determined the nature of trophic interactions; changes in climate at the end of the Pleistocene resulted in destabilization of the Arctic steppe–tundra ecosystem and altered the efficiency of energy transfer from one trophic level to the next. Disassembly of the Arctic steppe–tundra ecosystem at the beginning of the Holocene interglacial interval was probably a major factor in the widespread extinction of many species of large herbivorous mammals (Guthrie, 1982, 1984).

6.8 CONCLUSIONS

1. Stability, resilience and persistence of ecosystems are time-dependent functions that can be measured quantitatively using the paleoecological record.

2. Integrated paleoecological analyses of the coupled watershed-lake ecosystem yield new insights into long-term pattern and process at the ecosystem level. These relationships can be clarified through the study of the stratigraphic record of pollen grains representing upland vegetation, plant macrofossils of both upland and aquatic plants, sediment texture, chemistry, and deposition rates, algae including diatoms, and faunal remains including cladocera and molluscs.

3. During the late-Quaternary time interval, both natural environmental changes and anthropogenic activities have resulted in ecosystem-level effects at all trophic levels in both lakes and their surrounding watersheds.

4. Analysis of the potential of a given ecosystem for attaining 'steady state' or equilibrium conditions must take into account not only the history of environmental and anthropogenic changes in relation to the current status of the ecosystem, but also must consider the potential for future changes in

species availability, nutrient cycling and energy flow. Steady-state hypotheses must be conditioned by the realization that major climatic and ecosystem-level changes are inevitable over the long term.

5. Modern ecosystems do not provide modern analogues for all ecosystems that have existed in past times. Reconstruction of the structure and functioning of now-extinct ecosystems such as the late Pleistocene Arctic steppe–tundra yield tests of predictions made by ecosystem models that include trophic-level interactions of food webs.

7 Applications of Quaternary ecology to future global change

7.1 ISSUES

Future global climatic and environmental changes are emerging as one of the major research priorities of ecologists for the decade of the 1990s (Mooney, 1988). The international scientific community, through the International Council of Scientific Unions (ICSU), formally established the International Geosphere–Biosphere Program (IGBP) in 1986 in order to address imminent problems anticipated to result from global climatic change (Bolin *et al.*, 1986; Mooney, 1988). The overall objective of the IGBP is 'to describe and understand the interactive physical, chemical, and biological processes that regulate the total earth system, the unique environment that it provides for life, the changes that are occurring in this system, and the manner in which they are influenced by human activities' (National Research Council, 1988). One high priority research initiative involves the analysis of the history of the earth system to document long-term changes in atmospheric composition, climate, terrestrial and aquatic ecosystems, and human activities in order to validate and improve models of global environmental change (National Research Council, 1988). The mandate for studies in Quaternary ecology is to document population, community and ecosystem responses to rapid environmental changes that occurred in the past in order to provide insight into the rates and directions of biotic changes that may occur in the near future, as a consequence of climatic warming resulting from anthropogenic inputs of carbon dioxide (CO_2) and other 'greenhouse' gases to the atmosphere (National Research Council, 1986, 1988; M. Davis, 1988, 1990).

The Quaternary record of past changes in atmospheric CO_2, preserved within gas bubbles in polar ice (Delmas *et al.*, 1980; Neftel *et al.*, 1982, 1985), lends insight into the fundamental relationship between CO_2 concentrations and climate (Shackleton and Pisias, 1985). This polar-ice record establishes the long-term trajectory of prehistoric global changes in atmospheric carbon dioxide that has preceded the current amplification resulting from deforestation and burning of fossil fuels since the industrial revolution (Gammon *et al.*, 1985). Regional and global patterns of climatic change inferred from the Quaternary fossil record help identify the primary causes of climatic change and provide specific analogues for future climatic conditions. During former times of increased global temperatures, for example, the 'Hypsithermal Interval' of the middle-Holocene, from about 9000 to 4000 years ago (Wright, 1976), changes in position of ecotones between biomes, shifts in species distributions, and changes in plant and animal community structure may yield analogues for future changes to be anticipated within biotic communities. Ecotone shifts of the middle Holocene also have implications for changes in ecosystem function, for example, biomass and carbon storage (Billings, 1987; Solomon and Shugart, 1984). The late-Quaternary record of shifts in the distribution of individual taxa because of climatic warming is instrumental in evaluating potential consequences of future climatic warming to nature preserves established and managed to preserve biological diversity (M. Davis, 1988; Peters and Darling, 1985; Schonewald-Cox, 1988).

7.2 GLOBAL CLIMATIC CHANGE AND THE GREENHOUSE EFFECT

During the Quaternary Period, the Earth's climate system has been driven by long-term and periodic changes in solar radiation that generated global climatic change. The Earth's climate has oscillated between a glacial mode that favors the expansion of glaciers at middle and high latitudes and an interglacial mode that triggers the rapid disintegration of major glacial ice sheets. Each phase of cold climate favoring growth of glaciers lasts approximately 90 000 years and is followed by a relatively brief interglacial phase of warmth associated with a 10 000-year long interval during which glaciers melt and retreat. These global climatic changes are interpreted by most Quaternary scientists within the context of the prevailing paradigm of 'Milankovitch Cycles' (origin and development of this paradigm is discussed in Imbrie and Imbrie, 1979). The primary Milankovitch cycles have periodicities of 100 000 years, 41 000 years, and 21 000 years that correspond with the variation in both the amount and latitudinal distribution of solar radiation received by the Earth. Global climatic changes of the Quaternary thus have been the direct result of predictable, quantitative periodicities of the Earth's position as it wobbles on its rotational axis and orbits in an elliptical path around the Sun. The 100 000-year Milankovitch Cycle, generated by eccentricity variation in the Earth's elliptical orbit, amplifies feedbacks in the Earth's climate system, particularly in middle and high latitudes, and regulates the timing of glacial–interglacial cycles.

A plot of mean global temperature change (Fig. 7.1) illustrates a 5°C shift between glacial and interglacial climatic modes. Global temperature changes

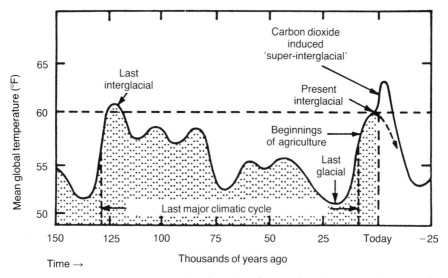

Figure 7.1 Global climate over the last glacial–interglacial cycle and projected for the next 25 000 years. A future cooling trend is predicted by the Milankovitch astronomical theory of the Ice Ages. However, this cooling trend is expected to be delayed over the next 2000 years because of a carbon dioxide-induced 'super-interglacial' interval of climatic warming triggered by historic human activities and combustion of fossil fuels. From Imbrie and Imbrie (1979).

over the last glacial–interglacial cycle (Fig. 7.1) are based upon oxygen isotope values measured in deep-sea records (Cline and Hays, 1976; CLIMAP, 1981; Shackleton *et al.*, 1983). Peak interglacial warmth occurred during the Sangamonian Interglacial interval between 130 000 and 120 000 years ago. The most recent phases of Wisconsinan continental glaciation expanded during the glacial episode dated between approximately 120 000 and 10 000 years ago. Peak warming (2°C warmer than today) occurred between 9000 and 4000 years ago during the Hypsithermal Interval of the Holocene Interglacial (Wright, 1976). Global cooling has occurred during the last several thousand years, indicating that the present interglacial period is nearing its end. Projections of climate for the next 25 000 years, in which global temperatures are predicted to decrease with the onset of another glaciation (Fig. 7.1), are based upon calculations of Milankovitch cycles. Alternatively, a major climatic warming of at least 2°C is proposed as a 'super-interglacial' (Fig. 7.1) that will last for at least the next millennium because of the anticipated build-up of CO_2 and the 'Greenhouse Effect' of trapping of infrared radiation within the atmosphere (Imbrie and Imbrie, 1979). Although the global average temperature is expected to increase by 2 to 6°C within the next century (Schneider, 1989), the increase will not be uniform over the surface of the earth; rather, it will be greatest at high

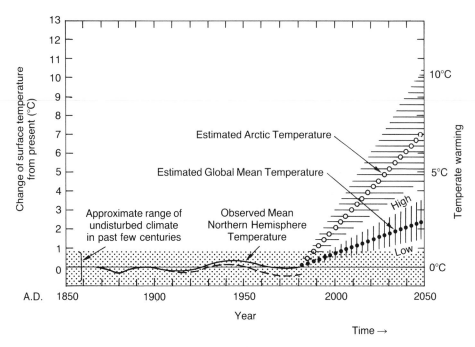

Figure 7.2   Observed record of mean Northern Hemisphere temperatures since A.D. 1850 and estimated increases in future global mean and arctic temperatures to A.D. 2050. The dashed line is an estimate of what the past temperature record would have been without increases in atmospheric carbon dioxide from the historic burning of fossil fuels. The shaded region denotes the 2°C range of global mean surface temperature over the past 1000 years. From Kellogg and Schware (1982).

latitudes, with arctic temperatures expected to rise by 5 to 10°C (Fig. 7.2; Kellogg and Schware, 1981, 1982).

Over the long term, values of atmospheric CO_2 (Fig. 7.3) have fluctuated within the range of 180 to 350 ppm, oscillating in step with major glacial–interglacial climatic changes during the past 150 000 years (Gammon *et al.*, 1985). Atmospheric concentrations of CO_2 for the past 40 000 years have been measured directly from air bubbles trapped within glacial ice of the Greenland and Antarctic ice caps (Delmas *et al.*, 1980; Neftel *et al.*, 1982, 1985). These studies demonstrate that full-glacial (20 000 years ago) concentrations of CO_2 were approximately 200 ppm and increased to an interglacial level of 280 ppm by 12 000 years ago, an increase of 80 ppm over 8000 years. Proxy measures of atmospheric CO_2 have been reconstructed for the past 150 000 years from isotopes of ^{13}C sampled from deep-sea sediment cores (Shackleton *et al.*, 1983). These CO_2 estimates derived from the paleo-oceanographic record generally verify and extend in time the glacial–interglacial trends recorded by the ice-core time series (Fig. 7.3). Atmospheric

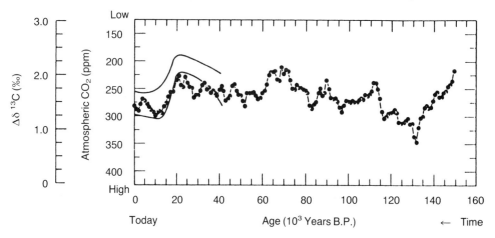

Figure 7.3 Changes in atmospheric carbon dioxide over the past 150 000 years interpreted from isotopic records in ocean sediment cores (dots) and compared with data from glacial ice cores for the past 40 000 years (heavy lines). From Gammon *et al.* (1985).

CO_2 values reached a peak value of 350 ppm 130 000 years ago during the Sangamonian Interglacial; they again reached relatively high values of between 270 and 280 ppm throughout the Holocene Interglacial (Fig. 7.3). Shackleton and Pisias (1985) and Genthon *et al.* (1987) suggest that long-term fluctuations in atmospheric CO_2 concentrations amplify as a positive feedback the global climatic changes of glacial–interglacial cycles that were initially triggered by Milankovitch cycles. Both ice-core studies (Neftel *et al.*, 1985) and tree-ring studies (Stuiver *et al.*, 1984) document that CO_2 concentrations have risen from a pre-industrial level of 280 ppm in A.D. 1800 to 345 ppm in 1985 (Gammon *et al.*, 1985). These carbon dioxide levels are projected to reach 600 ppm within the next 100 years (Schneider, 1989). Thus, CO_2 concentrations increased by 25% from peak glacial to interglacial times. These levels increased by another 25% within the last two centuries, and they are expected to double within the next 100 years (Schneider, 1989). The current rate of increase in CO_2 concentrations appears much greater than any prehistoric rates experienced in the Quaternary. In addition, the total CO_2 concentrations of 600 ppm expected by the year 2100 will be considerably higher than any atmospheric level occurring previously during the late-Quaternary interval.

7.3 PAST ANALOGUES FOR EFFECTS OF FUTURE GLOBAL WARMING

The mid-Holocene Hypsithermal Interval (Wright, 1976) may provide a partial analogue for the initial stages of global warming associated with the

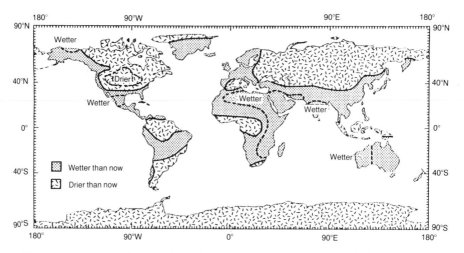

Figure 7.4 Future changes in soil moisture accompanying global warming as inferred from paleoclimatic reconstructions of previous warm, interglacial periods in the Quaternary, comparisons between recent warm and cold years or seasons, and a climate-simulation model. Where geographic areas of agreement occur among these sources, projected trends in soil moisture are indicated by either wetter or drier regions delineated by a dashed line. From Kellogg and Schware (1981).

Greenhouse Effect, including the effects of changes in regional geographical patterns of temperature and soil moisture on biotic responses. Kellogg and Schware (1981) mapped projected patterns of precipitation as departures from present-day conditions, primarily based upon paleoclimatic reconstructions for the Hypsithermal Interval, supplemented by modern meteorological data and climate-model simulations. Their map (Fig. 7.4) indicates that the mid-continental portions of North America, Europe, and the Soviet Union will be drier than today; a zone of increased precipitation will occur across Latin America, north Africa, and southeast Asia. Several kinds of biotic responses will occur to these shifts in climatic regions, including displacements of ecotones between major vegetation types and both elevational and latitudinal shifts in distribution of individual species. The late-Quaternary paleoecological record demonstrates the magnitude of these responses that can occur as the result of even a 2°C increase in global interglacial temperatures.

The classical example of an ecotone displacement resulting from mid-Holocene climatic warming and associated decreases in effective precipitation and soil moisture is that of changes in the location of the prairie–forest border in the mid-continental United States (Wright, 1968; 1976; Webb *et al.*, 1983). In presettlement times, prairie vegetation not only occupied the Great Plains region, but extended in a wedge known as the 'Prairie Peninsula' eastward as far as Illinois (King, 1981), with outliers of prairie communities found as far north as Michigan and Pennsylvania, and as far to the southeast as Kentucky and Arkansas (Transeau, 1935; for a discussion of the history of the concept of the Prairie Peninsula, see Stuckey, 1981). In the pollen record, regional prairie vegetation is characterized by ≥20% of prairie forb pollen, consisting of the sum of ragweed (*Ambrosia*), sage (*Artemisia*), other Compositae, and Chenopodiaceae/Amaranthaceae; pollen of grasses and sedges are not included in this sum because they may also have high values produced by local marsh vegetation around lake sites (Webb *et al.*, 1983). Maps with contours of pollen percentages ('isopoll maps') constructed from fossil pollen evidence derived from a number of lake sites throughout the upper Midwest region (Webb *et al.*, 1983) illustrate that the 20% isopoll contour marks the eastern edge of prairie and the western limit of forest. In the present interglacial, prairie first appeared in the Great Plains and western Great Lakes regions by 9000 years ago. The 'isochrone map' depicting changes in the location of the 20% contour for prairie forbs

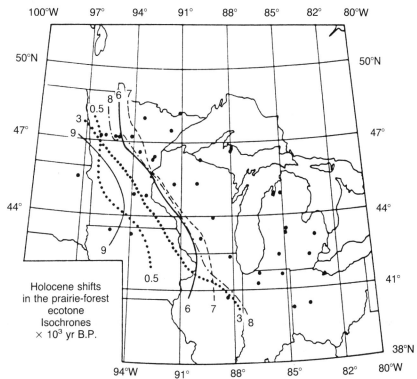

Figure 7.5 Map depicting Holocene changes in the position of the prairie/forest border in the Upper Midwest region of North America. Numbered lines locate the position of this ecotone (marked by the contour line for 20% prairie-forb pollen, with prairie located to the west and forest to the east of this line) at times indicated in thousands of years B.P. based upon fossil pollen data from paleoecological sites across the region (black dots). From Webb *et al.* (1983).

demonstrates that prairie reached its easternmost limit between 8000 and 7000 years ago (Fig. 7.5). Between 8000 and 6000 years ago, the prairie–forest border oscillated over about 1 to 3° of longitude. In the last 6000 years the ecotone between forest and prairie shifted westward by 2 to 3° of longitude (Webb *et al.*, 1983; Fig. 7.5). Using a multiple regression model for calibration of pollen and climate, Bart Bartlein *et al.* (1984) reconstructed Holocene changes in climate for the western Great Lakes region, extending from Manitoba south to Iowa and east to Indiana. Pollen-climate calibrations were applied to paleoecological sites throughout the region for the time span of 9000 years ago to 500 years ago. Between 9000 and 6000 years ago, reconstructed precipitation decreased by 10% to 25% through the northern Midwest. At 6000 years ago precipitation was <80% of modern values across Iowa, southern Minnesota, and southeastern Wisconsin. From 9000 to 6000 years ago an inferred increase in seasonal dominance of the Pacific Airmass (Wright, 1968; Bartlein *et al.*, 1984) resulted in an early-Holocene increase of 0.5 to 2°C in mean July temperatures. During the last 6000 years summer temperatures have generally decreased and precipitation has increased (Bartlein *et al.*, 1984). Thus, the middle-Holocene Hypsithermal Interval was associated with a regional climatic phenomenon that was not synchronous throughout its extent. Progressive changes in climate occurred as boundaries of airmasses changed in location in response to changes in global atmospheric circulation patterns. Vegetational changes along major ecotones such as the prairie–forest border were thus a 'time-transgressive' response (Wright, 1976) to intensifying gradients of not only temperature and precipitation, but of soil-moisture deficits and of disturbance regimes, particularly frequency of wildfire that determined the patchwork of prairie and forest communities across this ecotone (Grimm, 1983, 1984; Clark, 1988b, c; 1990).

Altitudinal shifts in vegetation in mountainous regions in response to regional expressions of global climatic change may be expected to reflect the individualistic nature of species tolerance thresholds (Davis *et al.*, 1980). Studies by Margaret Davis *et al.* (1980) and Ray Spear (1989) provide late-Quaternary records of vegetational change on an altitudinal transect through New England that ranges from near sea level in Connecticut up to 1542 meters at Lake of the Clouds in the White Mountains of northern New Hampshire. Postglacial vegetation development began in this region after retreat of glacial ice from the lowlands 15 000 years ago and from the mountain summits 13 000 years ago. During the late-glacial interval, tundra persisted for several thousand years in the lowlands and at intermediate elevations on mountain slopes, while barren periglacial desert characterized the summits of the White Mountains. Between 15 000 and 10 000 years ago, mean annual temperatures increased by 5° to 10°C; further climatic warming in the Holocene resulted in mean annual temperatures at least 2°C warmer than today between 9000 and 5000 years ago, during the Hypsithermal Interval. This estimate of past climate was based upon temperature lapse rates (0.6°C/100 m elevation) under the assumption that the correlation of modern species distributions with microclimate was applicable throughout the Holocene (M. Davis *et al.*, 1980; Spear, 1989). Both pollen influx and plant-macrofossil evidence indicate that between 11 000 and 9900 years ago, spruce and balsam fir (*Abies balsamea*) migrated up mountain slopes to an elevation of 1542 meters, near the present-day upper limit of black spruce (*Picea mariana*) and balsam fir krummholz (Spear, 1989). By 9000 years ago the subalpine forest of spruce and fir and the alpine tundra vegetation had established within elevational ranges comparable to those that they occupy today. White pine (*Pinus strobus*) migrated through the lowlands of southern New England 10 000 years ago and reached the White Mountains by 9000 years ago (Fig. 7.6; M. Davis *et al.*, 1980). By 8400 years ago, as documented by needles and pollen from Deer Lake Bog, white pine reached its modern upper elevational limit of 1325 meters, a vertical extension of 870 meters above its modern upper range limit of 455 meters in the White Mountains (Spear, 1989). White pine declined at its highest elevations after 6000 years ago (Spear, 1989), although macrofossil evidence demonstrates that it persisted at sites as high as 800 meters until the beginning of the Little Ice Age, approximately 500 years ago (M. Davis *et al.*, 1980). Based upon both pollen and macrofossils from the Mirror Lake site, eastern hemlock (*Tsuga canadensis*) arrived in the White Mountains by 8500 years ago; its populations were represented by macrofossils and pollen at an elevation of 1000 meters between 7000 and 5000 years ago (Fig. 7.6; M. Davis *et al.*, 1980). Today, the upper elevational limit of eastern hemlock is 610 meters; the Hypsithermal Interval distribution represented nearly a 400-meter altitudinal range extension. After 4800 years ago, the hemlock blight restricted the populations of eastern hemlock to an upper elevational limit of 650 meters. Little Ice Age climatic cooling resulted in a further restriction of eastern hemlock to its modern elevational range.

Thus, the increase in mean annual temperature associated with the middle-Holocene Hypsithermal Interval 9000 to 5000 years ago influenced the changes in distributions of coniferous tree species differentially in the White Mountains of New England. During the Hypsithermal Interval, climatic warming would have reduced the competitive capabilities of red and black spruce at their lower elevational limits, favoring the expansions of more temperate trees to higher elevations at the expense of spruce and fir. The differential elevational shifts in distribution of eastern hemlock and white pine illustrate individualistic responses to climatic change that resulted in compositional changes of mid-elevational conifer–northern hardwoods forests throughout the Holocene interval (Fig. 7.6). However, the ecotone between subalpine fir forest and alpine tundra communities persisted in place from 9000 years ago to the present. In order to raise the mean annual temperature to above 0°C, the threshold above which the geomorphic disturbance regime on the summits of the White Mountains would change

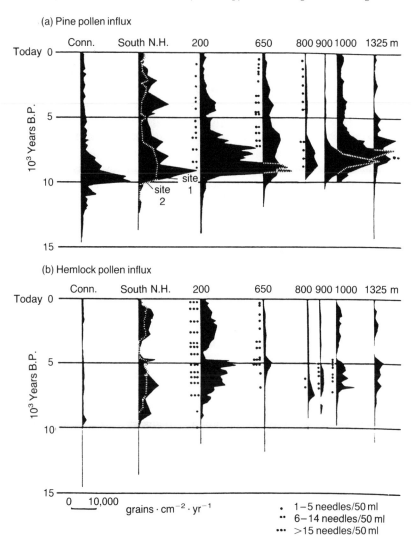

Figure 7.6   Pollen accumlation rates (influx) for (a) pine (*Pinus*) and (b) eastern hemlock (*Tsuga canadensis*) over the past 15 000 years at Rogers Lake, Connecticut, two lake sites in southern New Hampshire (silhouette and dashed line) and six paleoecological sites along an altitudinal transect from 200 meters to 1325 meters elevation in the White Mountains of New England. Dots indicate presence of macrofossils (needles), including identification of white pine (*Pinus strobus*) from sediments of the lake site at 1325 meters elevation at 8400 yr B.P.; (Spear, 1989). In the White Mountains today upper elevational limits are 455 meters for populations of white pine and 610 meters for eastern hemlock. Modified from Davis et al. (1980).

from discontinuous permafrost to a geomorphic environment in which trees could establish, climatic warming of the Hypsithermal Interval would have had to exceed 2.8°C (Spear, 1989). The presence of strong winds and wind exposure, resulting in extensive wind damage and desiccation of plants, are additional critical factors in determining the position of the treeline (Spear, 1989). Future responses of montane vegetation to climatic warming thus can be expected to reflect, in part, the immediate and individualistic elevational shifts of species according to their different life-history characteristics and climatic tolerances. These immediate responses will result in changes in community composition that have prehistoric analogues in the Hypsithermal Interval. However, for certain ecotones such as that between alpine forest and tundra, future climatic warming may have to exceed that previously experienced during the Holocene interval in order to evoke a vegetational response. Biotic change may be delayed until mean annual temperatures pass a critical geomorphic threshold, that is, the melting of the permafrost and a major decrease in intensity of frost-heaving of the soils, before the biota exhibits a measurable response.

7.4   CARBON BALANCE OF ALASKAN TUNDRA AND TAIGA ECOSYSTEMS

Dwight Billings (1987) illustrated the potential impacts of future global climatic change on a major Holocene sink for organic carbon in the high Arctic region of Alaska. He studied the paleoecological record of long-term dynamics of vegetation and carbon storage in tundra and taiga ecosystems and projected future ecosystem-level changes based upon phytotron simulation experiments. Billings (1987) emphasized that the arctic ecosystems contain approximately 27% of the total organic carbon sequestered within the global pool of soil carbon, with about 13.8% stored in tundra ecosystems, and another 13% stored within taiga ecosystems. Virtually 97% of the organic carbon within tundra is locked up in peat accumulations. This high percentage of organic carbon in soils is attributed to both the cold, wet climate that inhibits decomposition and the very shallow depth (20 to 25 cm) from soil surface to underlying permafrost.

Modern soil carbon pools, nutrient dynamics, and vegetation dynamics have been characterized for a major site within the International Biological Program (IBP) in the wet coastal tundra of the Alaskan North Slope near Point Barrow (Brown *et al.*, 1980). Paul Colinvaux (1964) published two pollen diagrams from sites at Barrow that provide a vegetational history spanning the past 14 000 years. The palynological evidence indicates the occurrence of sedge–grass–willow tundra with a cold, wet climate similar to that of today for the late-glacial and early-Holocene interval, radiocarbon dated between 14 000 and 9500 years ago. The onset of warmer climates between 9500 and

5000 years ago with local establishment of dwarf birch shrubs. The return to a cold, wet climate 5000 years ago has resulted in the persistence of the modern coastal tundra ecosystem. Billings (1987) interpreted the paleoecological data from other palynological sites as evidence for cyclic vegetational response of tundra communities to cyclic episodes of physical disturbance including development of thaw lakes and the generation of ice-wedge polygons through freeze-thaw processes. Billings (1987) concluded that the wet coastal tundra ecosystem has remained relatively unchanged for at least the last 5000 years. Modern rates of accumulation of soil carbon are approximately $41\,g\ C \cdot m^{-2} \cdot yr^{-1}$ (Chapin *et al.*, 1980). Billings (1987) considered that the carbon balance of this tundra ecosystem has resulted in the long-term capture of more carbon through photosynthesis than has been lost through respiration and decomposition. The excess has accumulated within the soil profile as peat and has been frozen at depth within the permafrost. Thus, the wet coastal tundra has provided a major carbon sink for most of the present Holocene interglacial.

A second study area that provides insights into the long-term dynamics of the carbon budget of the Arctic region is a taiga bog located in the Tanana River Valley near Fairbanks, Alaska (Billings, 1987). Today, the bog vegetation consists of a mosaic of open woodland with black spruce (*Picea mariana*), larch (*Larix laricina*) and paper birch (*Betula papyrifera*); a sedge meadow with cotton grass (*Eriophorum vaginatum*) and *Carex aquatilis*; and heath shrub communities including species of *Vaccinium*, *Andromeda polifolia*, *Ledum decumbens* subsp. *groenlandica* and dwarf birch (*Betula nana*). During the summer, the water table is at the land surface, with a thaw layer 20 to 60 cm deep overlying permafrost. Tom Ager (1983) summarized the late-Quaternary vegetational history of the Tanana Valley as follows: (1) herbaceous steppe–tundra persisted from full-glacial times until 14 000 years ago; (2) shrub tundra with dwarf birch, willow and ericads established from 14 000 to 11 500 years ago; (3) shrub tundra was replaced by scrub forest of *Populus* and willow from 11 500 to 9500 years ago; and (4) taiga forest of spruce, birch and alder has predominated for the last 9500 years. Ager (1983) considered that the boreal forest of the Tanana Valley established its modern composition by 6000 years ago. From an area of black spruce woodland within a bog site in the Tanana lowlands, a peat core of 192 cm length representing the past 4800 radiocarbon years has been studied for plant macrofossils and carbon content. The paleoecological data were used to determine a carbon budget for taiga soils spanning the mid- to late-Holocene interval (Billings, 1987). The peat stratigraphy (Fig, 7.7) reflected four episodes of vegetation development: (1) between 4800 and 4600 years ago, the site was occupied by a sedge meadow with low shrubs; (2) between 4600 and 3500 years ago, a spruce–birch forest produced abundant woody detritus; (3) between 3500 and 800 years ago, a sedge and low-shrub meadow with abundant tussocks of *Eriophorum vaginatum* established; and (4) in the last 800 years,

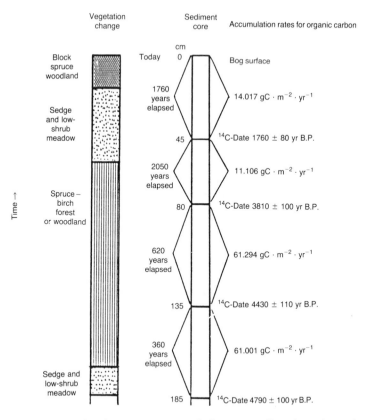

Figure 7.7   Diagram of Holocene vegetational changes, radiocarbon dates plotted against sediment depth, elapsed time between successive radiocarbon dates, and mean rates for accumulation of organic carbon (g $C \cdot m^{-2} \cdot yr^{-1}$) for the last 4800 years in a taiga bog at Fairbanks, Alaska, as interpreted from a sediment core. Modified from Billings (1987); reprinted with permission from Pergamon Press.

the site was occupied by a black spruce woodland with *Sphagnum* moss dominating the forest floor. Determined by loss-on-ignition on peat samples, values of organic carbon flux indicate that the most rapid accumulation of peat occurred between 4800 and 3800 years ago, averaging 61 g $C \cdot m^{-2} \cdot yr^{-1}$ (Fig. 7.7). The presence of charcoal particles in the peat samples from this middle-Holocene time indicates that the carbon accumulation rates are minimum values because some of the woody material was combusted in forest fires. After 3800 years ago, rates of organic carbon accumulation were reduced by over 75% to between 11 and 14 g $C \cdot m^{-2} \cdot yr^{-1}$ (Fig. 7.7), reflecting the vegetation shift to tussock sedge and low-shrub meadow and slow-growing spruce woodland. Overall, however, for at least the last 4800 years, the taiga of central Alaska has been a sink for organic carbon.

Short cores of peat and living surface vegetation were obtained from both tundra and taiga ecosystems to provide intact microcosms for analysis of carbon flux under elevated levels of atmospheric CO_2 and temperature in phytotron experiments (Billings, 1987). Simulations included increased summer temperatures, a doubling of atmospheric CO_2 concentrations, and lowered water tables expected to result from melting permafrost (Lachenbruch and Marshall, 1986). Altogether, the simulated environmental changes resulted in increased rates of CO_2 loss from the microcosms to the atmosphere. Under these conditions, subsurface peat accumulations would be exposed to increased oxidation and would become a major source of release of CO_2 to the atmosphere. Billings (1987) concluded that the integrity of arctic ecosystems is likely to break down in response to a greenhouse world with higher summer temperatures and carbon dioxide levels. These results are consistent with the projections of Bill Emanuel *et al.* (1985a, b) that, with the accelerated temperature increases associated with the Greenhouse Effect, destabilization of arctic ecosystems will result in a loss in area of circumboreal taiga by 37% and of arctic tundra by 32%. The importance of arctic ecosystems as a future source of atmospheric carbon, however, would not have been recognized in its proper perspective had the paleoecological record of this long-term, major carbon sink not been studied in order to provide the trajectory of change through the Holocene.

7.5 COMPUTER SIMULATIONS OF LONG-TERM CHANGES IN FOREST COMPOSITION AND BIOMASS

A variety of forest succession models has been produced to simulate stand development in successional time (Shugart, 1984). One of the most widely used models is FORET (Shugart and West, 1977), in which gap-phase dynamics of multi-storied forests are simulated through a model incorporating life-history characteristics of constituent species. This model and its numerous derivatives that have been developed for forest dynamics in boreal, temperate, and subtropical regions of the world (Shugart, 1984) incorporate environmental variables including climate, disturbance regime, nutrient availability, and soil moisture to drive changes in stand composition through differential reproduction, competition and mortality of tree populations. The models are generally validated using forest succession data from existing ecosystems in different geographical areas. Hank Shugart (1984) reviewed the assumptions and applications of forest-simulation models. These models have been applied to examine the possible range of future ecosystem dynamics as driven by environmental changes proposed for a greenhouse world, with differing combinations of native species, and under different scenarios of landscape management (Solomon and West, 1985, 1986).

Paleoecological studies have been interfaced with forest-stand simulation models in three ways: (1) to provide independent data sets for verification of long-term model simulations; (2) to simulate Hypsithermal Interval patterns of plant community and biomass changes; and (3) to yield a trajectory of forest dynamics from prehistoric into historic times to provide a baseline for projections of biotic responses to CO_2-induced climatic changes (Overpeck *et al.*, 1990).

Initial validation of the FORET model for application to long-term records of vegetational history was accomplished using the paleoecological record from Anderson Pond, Tennessee (Delcourt, 1979; Solomon *et al.*, 1980, 1981; Solomon and Shugart, 1984; Solomon and Webb, 1985; Solomon and Tharp, 1985). In a series of experiments, the pollen record for the past 16 000 years was compared with model output that incorporated 65 eastern North American tree species and several alternative interpretations of climatic history. The pollen record (Delcourt, 1979) documents that full-glacial vegetation was boreal forest of jack pine (*Pinus banksiana*) and spruce (*Picea*), with little representation of deciduous trees. Late-glacial climatic warming began by 16 500 years ago; deciduous forest taxa immigrated relatively rapidly, and late-glacial forests were composed of an admixture of boreal coniferous and temperate deciduous trees. By 12 500 years ago, oak (*Quercus*) became dominant, and populations of boreal trees became locally extinct as they migrated farther northward. After 9000 years ago, warm-temperate forests established. During the middle-Holocene Hypsithermal Interval, oak increased in importance relative to mesic deciduous trees such as sugar maple (*Acer saccharum*) and beech (*Fagus grandifolia*). Late-Holocene vegetation changes included a return to more mesic deciduous forest (Delcourt, 1979). The FORET model was used to simulate the last 16 000 years of forest history at Anderson Pond (Solomon *et al.*, 1980). Using inferred changes in growing-degree days to represent the temperature changes over the late-Quaternary interval, and allowing all 65 principal tree species of eastern North America to be continuously available for establishment in simulated plots on the Anderson Pond watershed, the FORET model was able to mimic both the timing and general nature of major changes in stand composition for this long-term record, including a rapid changeover from boreal to deciduous forest vegetation at about 13 500 years ago (Solomon *et al.*, 1980).

In a second study, variation in soil moisture was incorporated into the long-term simulations to test the effects of substrate and soil texture on forest stand composition. In this study, the FORET model was linked to the paleoecological records analyzed by Linda Brubaker (1975) in the Upper Peninsula of Michigan. Brubaker (1975) studied the contrasting pollen records from sites located within 12 kilometers of each other on sandy outwash soils (Yellow Dog Pond), silty loam till soils (Camp 11 Lake), and silty loam outwash soils (Lost Lake). Holocene temperature changes were mediated at these sites by differing water-holding capacities of the soils as well as by different recurrence intervals of fire (Brubaker, 1975). Early-Holocene forest communities were dominated by boreal species, with an open spruce woodland on the sandy outwash plain and codominance of spruce and jack pine on more mesic loamy soils. Middle-Holocene vegetation reflected the differential establishment of white pine (arriving 7800 years ago), a variety of deciduous species (immigrating generally after 7500 years ago), and invasion of eastern hemlock (arriving 5600 years ago). During the time of increased warming and drought severity of the mid-Holocene Hypsithermal Interval, an environmental gradient in both soil moisture and fire frequency resulted in jack pine barrens on sandy outwash plains, white pine-dominated forests on sandy loams, and a mixed deciduous forest on mesic silty loams (Brubaker, 1975). For the forest-stand simulation study, Al Solomon and Darrell West (1986) first developed a paleoclimate record to drive the model. They used pollen-climate calibrations based upon multivariate transfer functions to derive a time-series of estimates of changes in mean monthly growing-degree days and precipitation. Soil moisture was simulated by incorporating data on soil depth and potential water-holding capacity for each site. In the simulations, 73 tree species were potentially available in the simulated seed bank, but the sequence and timing of immigration of dominant tree taxa was controlled, based on the fossil pollen record. Three simulation runs were performed (Fig. 7.8) that combined climatic changes in conjunction with three different simulated soil moisture conditions. The first of these edaphic constraints on the model was expressed as 10% droughty days during the growing season at Camp 11 Lake, the second as 20% droughty days at Lost Lake, and the third as 30% droughty days at Yellow Dog Pond (Solomon and Shugart, 1984). From 10 350 to 7000 simulated years ago, modeled forest communities were dominated by birch, jack pine and red pine on dry and moist sites, with balsam fir and both black and white spruce occupying more mesic sites. For all simulated cases, mid-Holocene changes in simulated biomass projected population collapses of jack and red pine, moderate increases in populations of white pine and birch, and major population expansions of maple. For all three sites, total simulated biomass increased from 40 to 80 metric tons per hectare in the interval from 7000 to 5000 simulated years ago (Solomon and Shugart, 1984). For the xeric site (Yellow Dog Pond) and the intermediate site (Lost Lake), these simulations are implausible because fire disturbance regime, which would have diminished total biomass as well as affected stand composition at these sites, was not explicit included as an environmental factor in the forest-simulation model (Solomon and Shugart, 1984). Based on comparison with the paleoecological record, the simulated changes in biomass are most plausible for mesic silty loam sites such as occurred around Camp 11 Lake, both because soil droughtiness did not exceed tolerance thresholds for deciduous trees and because fires were not a frequent disturbance during the Holocene (Brubaker, 1975). The results of the forest-stand simulation model illustrate

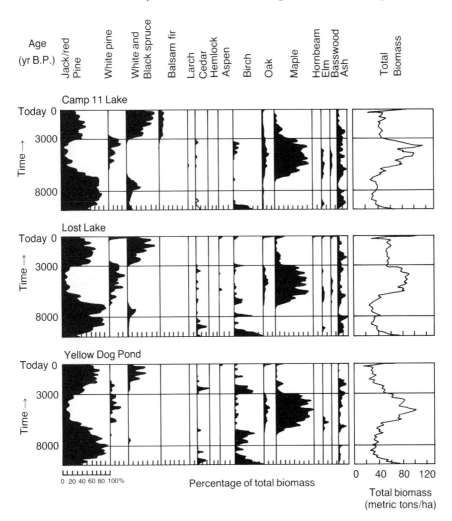

Figure 7.8 Simulated changes in biomass of principal trees in response to changing soil moisture and drought severity for the past 10 350 years for three sites in the Upper Peninsula of Michigan. From Solomon and Shugart (1984).

that changes in total biomass during the Holocene occurred on a time scale of millennia, reflecting not only the time required for establishment of deciduous trees on mesic sites but also the time required for progressive competitive exclusion of conifers by deciduous trees as forest systems responded to increased mid-Holocene temperatures.

After having been fine-tuned with paleoecological data, the forest-stand simulation model was applied to make projections of future changes in forest composition and biomass in three different geographic areas of eastern North

Simulated dynamics at sites in boreal, transition,
and deciduous forests

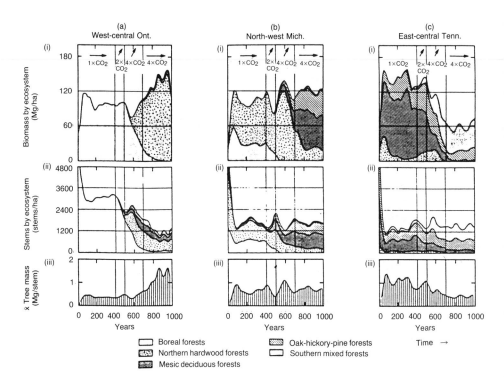

Figure 7.9 Simulated forest stand dynamics under changing climates resulting from increasing atmospheric carbon dioxide concentrations over a period of 1000 years at (a) a boreal forest site in west-central Ontario; (b) mixed conifer–northern hardwoods forest site in the Upper Peninsula of Michigan; and (c) deciduous forest site in east-central Tennessee. From Solomon and West (1986).

America (Solomon and West, 1986). The simulations of future CO_2 levels and associated climatic changes at each study site were based on climate simulations of Mitchell (1983) and Mitchell and Lupton (1984). Three hypothetical locations were evaluated (Fig. 7.9), including a boreal coniferous forest in west-central Ontario, a transitional boreal coniferous-temperate deciduous forest in northwestern Michigan, and a deciduous forest in east-central Tennessee. At all sites, the model simulated 1000 years of forest change, beginning 400 years ago and ending 600 years in the future. Biomass changes were simulated under conditions of a doubling of atmospheric CO_2 over the next 100 years (years 400 to 500 in the simulation) and then a quadrupling of atmospheric CO_2 between 100 and 300 years from now (years 500 to 700 in the simulation). At the location today occupied by boreal forest, summer and winter temperatures were projected to rise 2.5°C and 5°C, respectively between years 400 and 500 of the simulation, then to increase by another 2.5° and 4°C from years 500 to 700; precipitation was held constant throughout the 1000-year simulation. At the transitional site, seasonal temperatures increased by 2.5° and 3.5°C from years 400 to 500 and by 2.5° and 4°C from years 500 to 700; annual precipitation was held at today's levels until year 500, then decreased by 25% from year 500 to 700. At the deciduous forest site, summer and winter temperatures were increased by 3° and 2°C from year 400 to 500, then increased by 2° and 3°C from year 500 to 700; precipitation was held at today's levels until year 500, then decreased by 25%. These projected climatic changes have major consequences for overall biomass changes at the three sites (Fig. 7.9(i)). At the boreal location, stem numbers began to decrease as soon as climate began to warm, reflecting a shift to a mature forest with fewer, larger trees than before. Simulated stand biomass was unaffected until year 500 (100 years after the beginning of climatic warming). Upon doubling of CO_2 levels, however, total biomass declined for 50 to 75 years as large conifers died back. Subsequent establishment of hardwoods resulted in an increase in biomass that continued for several hundred years after climatic warming ended. At the transition site (Camp 11 Lake, Michigan), climatic warming caused an immediate and major dieback of mature boreal coniferous trees that were already under stress at their southern range limits. With continued warming, growth of cool-temperate hardwoods first increased in overall biomass, but then declined as their tolerance limits were exceeded and warm–temperate hardwoods began to replace them. At the deciduous forest site (Anderson Pond, Tennessee), simulated climatic and edaphic changes after year 500 were so great as to exceed the tolerance limits of tree growth, resulting in a permanent replacement of forest by open savanna. The results of this simulation study indicate that dramatic changes in vegetation composition and biomass may occur in the near future because of climatic changes associated with the Greenhouse Effect (Solomon and West, 1986).

Forest-stand simulation models have been verified independently and refined with long-term paleoecological records. The forest-stand models successfully mimic the timing and direction of compositional changes in prehistoric temperate and boreal forests. As Hypsithermal analogues for CO_2-induced climatic warming, prehistoric forest simulations emphasize three points: (1) a major global change in climate can be expected to trigger widespread biotic response that will take many centuries to millennia to approach steady-state conditions within the new climatic boundary conditions of a greenhouse earth; (2) in both magnitude and rate of climatic and biotic change, the Hypsithermal analogue can be expected to break down after the next 100 years as thresholds of species tolerances are exceeded; and (3) simulation projections of future biotic response will be plausible only to the extent that the models properly incorporate the response of native and exotic species to climatic change as well as to multiple stresses associated with changing disturbance regimes such as wildfire and environmental pollution including acid precipitation.

7.6 NATURE PRESERVES AND CONSERVATION OF BIOLOGICAL DIVERSITY

The Quaternary ecological record yields insights into the responses of individual species to environmental change and their consequences for plant community and ecosystem organization. These insights can be applied directly to address issues of conservation of biological diversity given projected future environmental changes. In a specific study designed to test the hypothesis that plant communities of the Great Lakes region will survive future global warming, Margaret Davis and Catherine Zabinski (1991) (see also M. Davis, 1988, 1989b; Roberts, 1989) examined the modern distributions of four principal cool–temperate tree species in light of their modern physiological tolerances and predicted the course of future changes in their distributions under two scenarios of climatic change. The first, less extreme projection of future climatic warming, the GISS model, developed by the Goddard Institute for Space Studies of the National Aeronautics and Space Administration (NASA) (Hansen *et al.*, 1983), forecast that with a doubling of atmospheric CO_2 by the year 2090, the Great Lakes region will experience an average warming of 4.5°C with slightly increased rainfall. A more extreme projection, made by the GFDL model of the National Oceanographic and Atmospheric Administration (NOAA) (Manabe and Weatherald, 1987), indicates an increase of 6.5°C for the region within the next 100 years, with a decrease in summer rainfall. M. Davis and Zabinski (1991) used maps of the current distributional ranges of beech (*Fagus grandifolia*), eastern hemlock (*Tsuga canadensis*), sugar maple (*Acer saccharum*) and yellow birch (*Betula allegheniensis*) (Little, 1971) as a point of departure for future shifts in their distributions. Present-day bioclimatic limits inferred from the mapped distributions were used to project future changes in distribution using estimates of mean January and July temperatures and of mean annual precipitation predicted by each of the two climate models. For every degree Celsius of mean annual temperature increase, a northward species range extension of 100 to 150 kilometers was inferred (Roberts, 1989). For all four species, range shifts of 500 to 1000 kilometers were predicted (450 to 675 kilometers for the GISS model; 650 to 975 kilometers for the GFDL model). Projected future distributional ranges for eastern hemlock (M. Davis, 1988) and beech (M. Davis and Zabinski, 1991) illustrate that both species may lose major populations over the southern extent of their current ranges, but that populations would be maintained within the central Great Lakes region and New England (GISS scenario), and the northern limits of these temperate tree species would be extended northward into Ontario and Quebec as far north as James Bay. Even in the extreme case (GFDL model), population centers would persist over time within the northern Appalachian Mountains in New England and Nova Scotia, providing a continuous seed source for migration to the north and northwest. Although future migrations would be individualistic, resulting in disassembly of present-day plant communities within the western Great Lakes region (M. Davis, 1988, 1989b), given sufficient time for migration, the principal species of today's mixed conifer–northern hardwoods communities would potentially occupy similar future ranges across southern Canada and the northeastern United States. Hence, forest communities that resemble those of today in composition and structure might or might not reassemble, depending upon whether the species are capable of establishing new populations at rates fast enough to keep pace with climatic change.

Potential rates of migration of these four tree species were examined by M. Davis and Zabinski (1991) and M. Davis (1988, 1989b). Based on analysis of the late-Quaternary record of species migrations in response to postglacial warming (M. Davis, 1981a), average Holocene migrational rates for eastern hemlock, beech, sugar maple, and yellow birch were all between 20 and 25 kilometers per century (M. Davis, 1988). Delcourt and Delcourt (1987a) calculated maximum rates of advance of tree populations during the portion of the Holocene that preceded the historic interval as 41 kilometers per century for maple, 58 kilometers per century for beech, 64 kilometers per century for birch, and 104 kilometers per century for eastern hemlock. The maximum observed rates of species migration for any North American tree taxon were 200 kilometers per century for spruce as it invaded sites across northwestern North America (Ritchie and MacDonald, 1986). The maximum Holocene rate of advance for any European tree species was 200 kilometers per century, achieved by alder (*Alnus*), a riparian taxon with broad climatic tolerances whose seed dispersal was facilitated by water transport in northward-flowing streams (Huntley and Birks, 1983). Margaret Davis (1988, 1989b) concluded that the rates of advance of temperate tree species during the late-Quaternary interval were much too slow to keep pace with the magnitude and rate of climatic change anticipated to occur within the next century (up to 6.5°C per hundred years). However, if the observed late-Quaternary rates of spread are minimum estimates because species were limited by the relative slowness of climatic changes rather than their inherent dispersal capabilities, then given several hundred to a thousand years for equilibration with future climatic change, the species could potentially accomplish the projected range shifts. In order for this to occur, both natural seed sources and suitable habitats (i.e. unmanaged nature preserves) would have to be available continuously. As an alternative, forest management practices in the future could include transplanting plant species to new locations outside their current ranges in order to facilitate reestablishment of

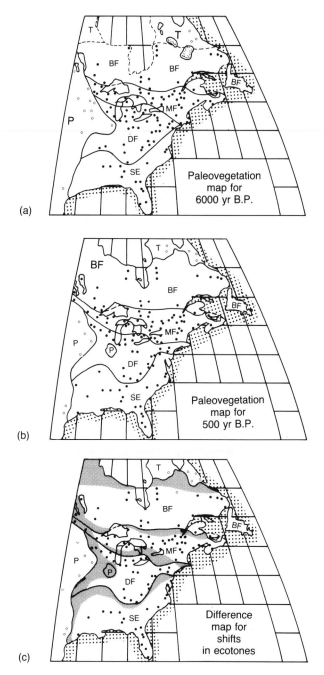

Figure 7.10 Paleovegetation maps for (a) 6000 yr B.P., (b) 500 yr B.P. and (c) a difference map showing shifts in ecotones in the past 6000 years in response to a change in mean annual temperature of about 2°C. Abbreviations for vegetation: T = tundra; P = prairie; BF = boreal forest; MF = mixed conifer–northern hardwoods forest; DF = deciduous forest; SE = southeastern evergreen forest. Modified from Delcourt and Delcourt (1987a).

the communities within the next century (M. Davis, 1988; M. Davis and Zabinski, 1991).

Although it is recognized that the responses of species of future climatic changes will be individualistic in nature, to the extent that the boundaries of world biomes correspond with fundamental positions of airmass boundaries (Bryson and Hare, 1974), the ecotones between biomes reflect zones of enhanced biological diversity at the limits of distribution of large numbers of species. These ecotones thus may be the areas to first register the effects of future global climatic change as tolerance thresholds are exceeded for species at the peripheries of their modern ranges (Delcourt and Delcourt, 1991). Differences in positions of ecotones between presettlement times (500 years ago) and peak Holocene interglacial conditions (e.g. 6000 years ago) represent biotic changes associated with a mean global temperature change of 2°C and yield a partial analogue for future warming of average global climate 2 to 6°C (Schneider, 1989), illustrating the geographical zones in which ecotone response to climatic warming may be first observed (Fig. 7.10). During the Holocene, substantial shifts of ecotones have occurred in four areas (Fig. 7.10): (1) the boundary between southeastern evergreen forest and deciduous forest; (2) the prairie–forest border; (3) the tundra–boreal forest transition; and (4) the region of mixed conifer–northern hardwoods forest, particularly across the western Great Lakes region (Delcourt and Delcourt, 1987a). Within the next 100 years (Schneider, 1989), coastal zones along both the Gulf of Mexico and the Atlantic Ocean, including much of the lowland area of Florida, will experience a rise in sea level of up to 6 meters that will cause major shifts in the ecotone between coastal marshes and upland vegetation (Spackman *et al.*, 1966; M. Davis, 1988).

Even within central areas of biomes, populations or entire species may become extinct (for example, those populations of beech that today occur within the eastern deciduous forest), resulting in a depauperization of communities. Location of nature reserves with the intent to preserve biological diversity therefore must take into account: (1) the late-Quaternary history of assembly of biotic communities and the evidence for ecosystem resilience to environmental change; (2) proximity to major ecotonal boundaries; (3) past, present, and potential future gradients in species abundance and diversity with distance from those ecotones; (4) the minimum critical area for maintaining species populations in a particular preserve; and (5) the spacing of preserves along potential migrational corridors in order to maximize the success of dispersal and establishment of migrating species (Peters and Darling, 1985; M. Davis, 1988; Cohn, 1989; M. Davis and Zabinski, 1991; Delcourt and Delcourt, 1987a, 1991). For example (Fig. 7.11; Peters and Darling, 1985), placement of a nature 'preserve' near the southern limits of distribution of a number of species today may result in loss of that area as a preserve for those particular species as they contract their ranges to the north and are replaced by immigrating species from the south. Placement

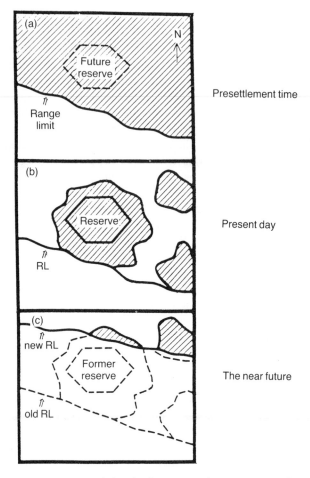

Figure 7.11   Schematic representation of the displacement of species range limits beyond the boundaries of a biological reserve under changing climatic conditions. The shaded pattern indicates (a) distribution of the species before human distur- bance or climate change (RL = southern range limit); (b) fragmentation of the species distribution because of human disturbance; (c) movement of species range limits because of climatic warming. From Peters and Darling (1985); copyright 1985 by the American Institute of Biological Sciences.

of a 'preserve' near the northern limits of distribution of a group of species (for example, in mixed conifer–northern hardwoods forests of northern Maine) may help to maintain viable, natural seed sources for future coloni- zation of new potential ranges of species. Location of preserves in regions of high elevational relief, such as in the Appalachian or Rocky Mountains, would help to ensure maximum habitat diversity in the future in order for species to shift their altitudinal limits individualistically in response to climatic change (as observed in the fossil record in the eastern Grand Canyon (Cole, 1982, 1985) and in the White Mountains of New Hampshire (M. Davis *et al.*, 1980; Spear, 1989)). Successful dispersal and establishment of beech across the Prairie Peninsula during the Hypsithermal Interval (S. Webb, 1986) documents the prehistoric migration of a tree species with dispersal jumps of up to 130 kilometers between forest islands within a matrix of prairie vegetation. Thus, with an array of forest preserves situated at distances of no more than 100 kilometers within an agricultural landscape, dispersal and establishment of tree populations may remain viable. New nature preserves must be created at tropical, temperate, boreal, and arctic latitudes in order to ensure long-term maintenance of global patterns in biodiversity (M. Davis and Zabinski, 1991).

7.7 CONCLUSIONS

1. The middle-Holocene Hypsithermal Interval provides a partial analogue for biotic changes that may be associated with future global climatic warming resulting from the Greenhouse Effect. The Hypsithermal Interval provides valuable precedents for ecotone shifts, individualistic behavior of migrating species, and changes in long-term carbon storage of ecosystems. The Hypsithermal analogue is most useful in anticipating the nature and geographic patterns of biotic response to the initial 2°C warming of mean global temperature in the first stage of the CO_2-induced 'super-interglacial'.

2. Both the magnitude and rate of climatic change in the next 100 years are expected to exceed those observed in the late-Quaternary interval, resulting in the differential range shifts of species as their southern populations become extinct and as they spread northward into new potential ranges across the northern temperate zone. Present knowledge of species tolerances and interactions is insufficient to predict with certainty their population dynamics in response to conditions that will be beyond those that have existed in the late-Quaternary interval.

3. In the absence of direct human intervention, modern biotic communities may disassemble with differential loss of species as they experience either local or global extinctions. These communities may possibly reassemble in new geographical areas, but only after a lag of several millennia associated with differential migrational trajectories of species as constrained by life-history strategies, biological inertia, locations of seed sources and available dispersal routes.

4. In North America, paleoecological studies for the Hypsithermal Interval emphasize the resilience of natural ecosystems in the absence of major human modification of the landscape. Today, human activities are producing new and unique combinations of multiple stresses through forest

fragmentation and environmental pollution that compound the difficulty in predicting future biotic responses.

5. The Quaternary ecological record identifies the geographical areas that are most vulnerable to future ecosystem changes and thus provides important information to ecological managers in their selection of sites for ecological preserves. Choices facing conservationists include identifying a series of preserves as 'stepping stones' or 'safe-sites' to facilitate species migrations, or, alternatively, focusing attention on preserving examples of contemporary communities and of populations of rare and endangered species.

References

Aaby, B. (1976) Cyclic climatic variations in climate over the past 5500 yr reflected in raised bogs. *Nature*, **263**, 281–5.

Ager, T. A. (1983) Holocene vegetational history of Alaska, in *Late Quaternary Environments of the United States*, Vol. 2, *The Holocene* (ed. H. E. Wright, Jr.), University of Minnesota Press, Minneapolis, pp. 128–41.

Allen, T. H. F. and Starr, T. B. (1982) *Hierarchy: Perspectives for Ecological Complexity*, University of Chicago Press, Chicago.

Allison, T. D., Moeller, R. E. and Davis, M. B. (1986) Pollen in laminated sediments provides evidence for a mid-Holocene forest pathogen outbreak. *Ecology*, **67**, 1101–5.

Ammann, B. (1986) Litho- and palynostratigraphy at Lobsigensee: evidences for trophic changes during the Holocene. *Hydrobiologia*, **143**, 301–7.

Ammann, B. (1988) Palynological evidence of prehistoric anthropogenic forest changes on the Swiss Plateau, in *The Cultural Landscape – Past, Present and Future* (eds H. H. Birks, H. J. B. Birks, P. E. Kaland and D. Moe), Cambridge University Press, Cambridge, pp. 289–302.

Ammann, B. (1989a) Response times in bio- and isotope-stratigraphies to Late-Glacial climatic shifts – an example from lake deposits. *Ecologae geol. Helv.*, **82**, 5–12.

Ammann, B. (1989b) Late-Quaternary palynology at Lobsigensee: regional vegetation history and local lake development. *Dissertationes Botanicae*, Band 137, J. Cramer, Berlin.

Ammann, B. and Tobolski, K. (1983) Vegetational development during the Late-Wurm at Lobsigensee (Swiss Plateau). *Rev. Paleobiol.*, **2**, 163–80.

Ammann, B., Chaix, L., Eicher, U., Elias, S. A., Gaillard, M.-J., Hofmann, W., Siegenthaler, U., Tobolski, K. and Wilkinson, B. (1983) Vegetation, insects, molluscs and stable isotopes from Late-Wurm deposits at Lobsigensee (Swiss Plateau). *Rev. Paleobiol.*, **2**, 221–7.

Andersen, S. T. (1966) Interglacial vegetation succession and lake development in Denmark. *Palaeobotanist*, **15**, 117–27.

Andersen, S. T. (1984) Forests at Lovenholm, Djursland, Denmark, at present and in the past. *Det Kingelige Danske Videnskabernes Selskab Biologiske Skrifter*, **24**, 1–208.

Anderson, R. S., Davis, R. B., Miller, N. G. and Stuckenrath, R. (1986) History of late- and post-glacial vegetation and disturbance around Upper South Branch Pond, northern Maine. *Can. J. Bot.*, **64**, 1977–86.

Anderson, T. W. (1974) The chestnut pollen decline as a time horizon in lake sediments in eastern North America. *Can. J. Earth Sci.*, **11**, 678–85.

Ashworth, A. C., Schwert, D. P., Watts, W. A. and Wright, H. E., Jr. (1981) Plant and insect fossils at Norwood in south-central Minnesota: a record of late-glacial succession. *Quat. Res.*, **16**, 66–79.

Balsam, W. (1981) Late Quaternary sedimentation in the western North Atlantic:

stratigraphy and paleoceanography. *Palaeogeogr. Palaeoclimatol. Palaeoecol.*, **35**, 215–40.

Barber, K. E. (1981) *Peat Stratigraphy and Climatic Change: a Palaeoecological Test of the Theory of Cyclic Peat Bog Regeneration*, A. A. Balkema, Rotterdam.

Barbour, M. G., Burk, J. H. and Pitts, W. D. (1987) *Terrestrial Plant Ecology*, 2nd edn, Benjamin/Cummings, Menlo Park, CA.

Barker, G. (1985) *Prehistoric Farming in Europe*, Cambridge University Press, Cambridge.

Barnosky, C. W. (1987) Response of vegetation to climatic changes of different duration in the Late Neogene. *Trends Ecol. Evol.*, **2**, 247–50.

Bartlein, P. J. (1988) Late-Tertiary and Quaternary palaeoenvironments, in *Vegetation History* (eds B. Huntley and T. Webb III), Kluwer, Dordrecht, pp. 113–52.

Bartlein, P. J., Webb, T. III and Fleri, E. (1984) Holocene climatic change in the northern Midwest: pollen-derived estimates. *Quat. Res.*, **22**, 361–74.

Bartlein, P. J., Prentice, I. C. and Webb, T. III (1986) Climatic response surfaces from pollen data for some eastern North American taxa. *J. Biogeogr.*, **13**, 35–57.

Begon, M. and Mortimer, M. (1982) *Population Ecology*, Blackwell Scientific, Oxford.

Behre, K.-E. (editor) (1986) *Anthropogenic Indicators in Pollen Diagrams*, A. A. Balkema, Rotterdam.

Behre, K.-E. (1989) Biostratigraphy of the last glacial period in Europe. *Quat. Sci. Rev.*, **8**, 25–44.

Bennett, K. D. (1983) Postglacial population expansion of forest trees in Norfolk, UK. *Nature*, **303**, 164–7.

Bennett, K. D. (1985) The spread of *Fagus grandifolia* across eastern North America during the last 18 000 years. *J. Biogeogr.*, **12**, 147–64.

Bennett, K. D. (1988) Post-glacial vegetation history: ecological considerations, in *Vegetation History* (eds B. Huntley and T. Webb III), Kluwer, Dordrecht, pp. 699–724.

Bennett, K. D. and Lamb, H. F. (1988) Holocene pollen sequences as a record of competitive interactions among tree populations. *Trends Ecol. Evol.*, **3**, 141–4.

Benninghoff, W. S. (1964) The Prairie Peninsula as a filter barrier to postglacial plant migration. *Proc. Indiana Acad. Sci.*, **72**, 116–24.

Berglund, B. E. (ed.) (1986) *Handbook of Holocene Palaeoecology and Palaeohydrology*, Wiley and Sons, New York.

Bernabo, J. C. (1981) Quantitative estimates of temperature changes over the last 2700 years in Michigan based on pollen data. *Quat. Res.*, **15**, 143–59.

Bernabo, J. C. and Webb, T. III. (1977) Changing patterns in the Holocene pollen record of northeastern North America: a mapped summary. *Quat. Res.*, **8**, 64–96.

Billings, W. D. (1987) Carbon balance of Alaskan tundra and taiga ecosystems: past, present and future. *Quat. Sci. Rev.*, **6**, 165–77.

Binford, M. W., Brenner, M., Whitmore, T. J., Higuera-Gundy, A., Deevey, E. S. and Leyden, B. (1987) Ecosystems, paleoecology and human disturbance in subtropical and tropical America. *Quat. Sci. Rev.*, **6**, 115–28.

Birks, H. H. (1980) Plant macrofossils in Quaternary lake sediments. *Ergebnisse der Limnologie*, **15**, 1–60.

Birks, H. J. B. (1976) Late-Wisconsinan vegetational history at Wolf Creek, central Minnesota. *Ecol. Monogr.*, **46**, 395–429.

Birks, H. J. B. (1980a) The present flora and vegetation of the moraines of the Klutlan Glacier, Yukon Territory, Canada: a study in plant succession. *Quat. Res.*, **14**, 60–86.

Birks, H. J. B. (1980b) Modern pollen assemblages and vegetational history of the moraines of the Klutlan Glacier and its surroundings, Yukon Territory, Canada. *Quat. Res.*, **14**, 101–29.

Birks, H. J. B. (1981a) Late Wisconsin vegetational and climatic history at Kylen Lake, northeastern Minnesota. *Quat. Res.*, **16**, 322–55.

Birks, H. J. B. (1981b) The use of pollen analysis in the reconstruction of past climates: a review, in *Climate and History* (eds T. M. L. Wigley, M. J. Ingram and G. Farmer), Cambridge University Press, Cambridge, pp. 111–38.

Birks, H. J. B. (1986) Late-Quaternary biotic changes in terrestrial and lacustrine environments, with particular reference to north-west Europe, in *Handbook of Holocene Palaeoecology and Palaeohydrology* (ed. B. E. Berglund), John Wiley and Sons, New York, pp. 3–65.

Birks, H. J. B. (1989) Holocene isochrone maps and patterns of tree-spreading in the British Isles. *J. Biogeogr.*, **16**, 503–40.

Birks, H. J. B. and Birks, H. H. (1980) *Quaternary Palaeoecology*, University Park Press, Baltimore.

Birks, H. J. B. and Gordon, A. D. (1985) *Numerical Methods in Quaternary Pollen Analysis*, Academic Press, New York.

Bliss, L. C. and Richards, J. H. (1982) Present-day arctic vegetation and ecosystems as a predictive tool for the Arctic-Steppe Mammoth biome, in *Paleoecology of Beringia* (eds D. M. Hopkins, J. V. Matthews, Jr., C. E. Schweger, and S. B. Young), Academic Press, London, pp. 241–57.

Bolin, B., Doos, B. R., Jager, J. and Warrick, R. (eds) (1986) *SCOPE 29: Greenhouse Effect, Climatic Change, and Ecosystems*. John Wiley and Sons, Chichester.

Bormann, F. H. and Likens, G. E. (1979) *Pattern and Process in a Forested Ecosystem: Disturbance, Development and the Steady State Based on the Hubbard Brook Ecosystem Study*, Springer-Verlag, New York.

Bradshaw, R. H. W. and Miller, N. G. (1988) Recent successional processes investigated by pollen analysis of closed-canopy sites. *Vegetatio*, **76**, 45–54.

Braun, E. L. (1950) *Deciduous Forests of Eastern North America*, Hafner, New York. (Reprinted in 1974.)

Brown, J., Everett, K. R., Webber, P. J., MacLean, S. F., Jr and Murray, D. F. (1980) The coastal tundra at Barrow, in *An Arctic Ecosystem: the Coastal Tundra at Barrow, Alaska* (eds J. Brown, P. C. Miller, L. L. Tieszen and F. L. Bunnell), Dowden, Hutchinson, and Ross, Stroudsburg, PA, pp. 1–29.

Brubaker, L. B. (1975) Postglacial forest patterns associated with till and outwash in northcentral Upper Michigan. *Quat. Res.*, **5**, 499–527.

Brugam, R. B. (1978a) Pollen indicators of land-use change in southern Connecticut. *Quat. Res.*, **9**, 349–62.

Brugam, R. B. (1978b) Human disturbance and the historical development of Linsley Pond. *Ecology*, **59**, 19–36.

Brugam, R. B. (1980) Post-glacial diatom stratigraphy of Kirchner Marsh, Minnesota. *Quat. Res.*, **13**, 133–46.

Brugam, R. B. (1983) Holocene paleolimnology, in *Late-Quaternary Environments of the United States*, Vol. 2, *The Holocene* (ed. H. E. Wright, Jr.), University of Minnesota Press, Minneapolis, pp. 208–21.

Bryson, R. A. (1966) Air masses, streamlines, and the boreal forest. *Geogr. Bull.*, **8**, 228–69.

Bryson, R. A. and Hare, F. K. (1974) The climates of North America, in *Climates of North America* (eds R. A. Bryson and F. K. Hare), Elsevier, New York, pp. 1–47.

Bryson, R. A. and Wendland, W. M. (1967) Tentative climatic patterns of some late glacial and postglacial episodes in central North America, in *Life, Land, and Water, Proceedings of the 1966 Conference on Environmental Studies of the Glacial*

Lake Agassiz Region (ed. W. J. Mayer-Oakes), University of Manitoba Press, Winnipeg, pp. 271–98.

Bryson, R. A., Baerreis, D. A. and Wendland, W. M. (1970) The character of late-glacial and post-glacial climatic changes, in *Pleistocene and Recent Environments of the Central Great Plains* (eds W. Dort, Jr. and J. K. Jones, Jr.), University of Kansas Press, Lawrence, pp. 53–74.

Burgess, R. L. and Sharpe, D. M. (1981) *Forest Island Dynamics in Man-Dominated Landscapes*, Springer-Verlag, New York.

Burke, M. J., George, M. F. and Bryant, R. G. (1975) Water in plant tissues and frost hardiness, in *Water Relations of Foods* (ed. R. B. Duckworth), Academic Press, New York, pp. 111–35.

Burke, M. J., Gusta, L. V., Quamme, H. A., Weiser, C. J. and Li, P. H. (1976). Freezing and injury in plants. *Ann. Rev. Plant Physiol.*, **27**, 507–28.

Carter, R. N. and Prince, S. D. (1981) Epidemic models used to explain biogeographical distribution limits. *Nature*, **293**, 644–5.

Chabot, B. F. and Mooney, H. A. (1985) *Physiological Ecology of North American Plant Communities*, Chapman and Hall, New York.

Chaix, L. (1983) Malacofauna from the Late-Glacial deposits of Lobsigensee (Swiss Plateau). *Rev. Paleobiol.*, **2**, 211–6.

Chapin, F. S., III, Miller, P. C., Billings, W. D. and Coyne, P. I. (1980) Carbon and nutrient budgets and their control in coastal tundra, in *An Arctic Ecosystem: the Coastal Tundra at Barrow, Alaska* (eds J. Brown, P. C. Miller, L. L. Tieszen and F. L. Bunnell), Dowden, Hutchinson, and Ross, Stroudsburg, PA, pp. 458–82.

Charles, D. F. (1985) Relationships between surface sediment diatom assemblages and lakewater characteristics in Adirondack lakes. *Ecology*, **66**, 994–1011.

Charles, D. F., Whitehead, D. R., Engstrom, D. R., Fry, B. D., Hites, R. A., Norton, S. A., Owen, J. S., Roll, L. A., Schindler, S. C., Smol, J. P., Uutala, A. J., White, J. R. and Wise, R. J. (1987) Paleolimnological evidence for recent acidification of Big Moose Lake, Adirondack Mountains, N. Y. (USA). *Biogeochemistry*, **3**, 267–96.

Clark, J. S. (1986) Dynamism in the barrier-beach vegetation of Great South Beach, New York. *Ecol. Monogr.*, **56**, 97–126.

Clark, J. S. (1988a) Particle motion and the theory of charcoal analysis: source area, transport, deposition, and sampling. *Quat. Res.*, **30**, 67–80.

Clark, J. S. (1988b) Stratigraphic charcoal analysis on petrographic thin sections: application to fire history in northwestern Minnesota. *Quat. Res.*, **30**, 81–91.

Clark, J. S. (1988c) Effect of climate change on fire regimes in northwestern Minnesota. *Nature*, **334**, 233–5.

Clark, J. S. (1989a) Ecological disturbance as a renewal process: theory and application to fire history. *Oikos*, **56**, 17–30.

Clark, J. S. (1989b) Effects of long-term water balances on fire regime, north-western Minnesota. *J. Ecol.*, **77**, 989–1004.

Clark, J. S. (1990) Fire and climate change during the last 750 years in northwestern Minnesota. *Ecol. Monogr.*, **60**, 135–69.

Clark, J. S. and Patterson, W. A. III (1985) The development of a tidal marsh: upland and oceanic influences. *Ecol. Monogr.*, **55**, 189–217.

Clements, F. E. (1916) Plant succession: an analysis of the development of vegetation. Publication 242, Carnegie Institute of Washington, Washington, D.C.

CLIMAP Project Members (1981) Seasonal reconstructions of the Earth's surface at the last glacial maximum. *Geol. Soc. Amer. Map Ser.*, MC-36.

Cline, R. M. and Hays, J. D. (eds) (1976) CLIMAP: investigations of Late

Quaternary paleoceanography and paleoclimatology. *Geol. Soc. Amer. Mem.*, **145**, 1–464.

Cloud, P. (1976) Beginnings of biospheric evolution and their biogeochemical consequences. *Paleobiology*, **2**, 351–87.

COHMAP (1988) Climatic changes of the last 18 000 years: observations and model simulations. *Science*, **241**, 1043–52.

Cohn, J. P. (1989) Gauging the biological impacts of the greenhouse effect. *Bioscience*, **39**, 142–6.

Cole, K. (1982) Late Quaternary zonation of vegetation in the eastern Grand Canyon. *Science*, **217**, 1142–45.

Cole, K. (1985) Past rates of change, species richness, and a model of vegetational inertia in the Grand Canyon, Arizona. *Amer. Nat.*, **125**, 289–303.

Cole, K. (1986) In defense of inertia. *Amer. Nat.*, **127**, 727–8.

Colinvaux, P. A. (1964) Origin of ice ages: pollen evidence from arctic Alaska. *Science*, **145**, 707–8.

Colinvaux, P. A. (1987) Amazon diversity in light of the paleoecological record. *Quat. Sci. Rev.*, **6**, 93–114.

Cowles, H. C. (1899) The ecological relations of the vegetation on the sand dunes of Lake Michigan. *Bot. Gaz.*, **27**, 95–117, 167–202, 281–308, 361–91.

Craig, A. J. (1972) Pollen influx to laminated sediments: a pollen diagram from northeastern Minnesota. *Ecology*, **53**, 46–57.

Crosby, A. W. (1986) *Ecological Imperialism: The Biological Expansion of Europe, 900–1900.* Cambridge University Press, Cambridge.

Curtis, J. T. (1959) *The Vegetation of Wisconsin: an Ordination of Plant Communities*, University of Wisconsin Press, Madison.

Cushing, E. J. (1964) Redeposited pollen in late-Wisconsin pollen spectra from east-central Minnesota. *Amer. J. Sci.*, **262**, 1075–88.

Cwynar, L. C. and MacDonald, G. M. (1987) Geographical variation of lodgepole pine in relation to population history. *Amer. Nat.*, **129**, 463–9.

Cwynar, L. C. and Ritchie, J. C. (1980) Arctic steppe–tundra: a Yukon perspective. *Science*, **208**, 1375–7.

Davis, M. B. (1963) On the theory of pollen analysis. *Amer. J. Sci.*, **261**, 897–912.

Davis, M. B. (1969a) Pollen accumulation rates at Rogers Lake, Connecticut, during late- and postglacial time. *Rev. Palaeobot. Palynol.*, **2**, 219–30.

Davis, M. B. (1969b) Palynology and environmental history during the Quaternary Period. *Amer. Scientist*, **57**, 317–32.

Davis, M. B. (1973) Redeposition of pollen grains in lake sediment. *Limnol. Oceanogr.*, **18**, 44–52.

Davis, M. B. (1976) Pleistocene biogeography of temperate deciduous forests. *Geoscience and Man*, **13**, 13–26.

Davis, M. B. (1978) Climatic interpretation of pollen in Quaternary sediments, in *Biology and Quaternary Environments* (eds D. Walker and J. C. Guppy), Australian Academy of Science, Canberra, ACT, pp. 35–51.

Davis, M. B. (1981a) Quaternary history and the stability of forest communities, in *Forest Succession: Concepts and Application* (eds D. C. West, H. H. Shugart and D. B. Botkin), Springer-Verlag, New York, pp. 132–53.

Davis, M. B. (1981b) Outbreaks of forest pathogens in Quaternary history. *Proc. IV Int. Palynol. Conf. Lucknow* (1976–1977), **3**, 216–27.

Davis, M. B. (1983) Quaternary history of deciduous forests of eastern North America and Europe. *Ann. Mo. Bot. Gard.*, **70**, 550–63.

Davis, M. B. (1985) History of the vegetation on the Mirror Lake watershed, in *An*

Ecosystem Approach of Aquatic Ecology, Mirror Lake and its Environment (ed. G. E. Likens), Springer-Verlag, New York, pp. 53–65.

Davis, M. B. (1986) Climatic instability, time lags, and community disequilibrium, in *Community Ecology* (eds J. Diamond and T. J. Case), Harper and Row, New York, pp. 269–84.

Davis, M. B. (coordinator) (1988) Ecological systems and dynamics, in *Toward an Understanding of Global Change: Initial Priorities for U.S. Contributions to the International Geosphere-Biosphere Program* (ed. National Research Council), National Academy Press, Washington, DC, pp. 69–106.

Davis, M. B. (1989a) Retrospective studies, in *Long-Term Studies in Ecology* (ed. G. E. Likens), Springer-Verlag, New York, pp. 71–89.

Davis, M. B. (1989b) Lags in vegetation response to greenhouse warming. *Climatic Change*, **15**, 75–82.

Davis, M. B. (coordinator) (1990) Special Issue: biology and palaeobiology of global climate change. *Trends Ecol. Evol.*, **5**, 269–322.

Davis, M. B. and Botkin, D. B. (1985) Sensitivity of cool-temperate forests and their fossil pollen record to rapid temperature change. *Quat. Res.*, **23**, 327–340.

Davis, M. B. and Ford, M. S. (J.) (1982) Sediment focusing in Mirror Lake, New Hampshire. *Limnol. Oceanogr.*, **27**, 147–50.

Davis, M. B. and Zabinski, C. (1991) Changes in geographical range resulting from greenhouse warming – effects of biodiversity in forests, in *Consequences of Greenhouse Warming to Biodiversity* (eds R. L. Peters and T. Lovejoy), Yale University Press (in press).

Davis, M. B., Brubaker, L. B. and Beiswenger, J. M. (1971) Pollen grains in lake sediments: pollen percentages in surface sediments from southern Michigan. *Quat. Res.*, **1**, 450–67.

Davis, M. B., Brubaker, L. B. and Webb, T., III. (1973) Calibration of absolute pollen influx, in *Quaternary Plant Ecology* (eds H. J. B. Birks and R. G. West), Wiley and Sons, New York, pp. 9–25.

Davis, M. B., Ford, M. S. (J.) and Moeller, R. E. (1985a) Paleolimnology: sedimentation, in *An Ecosystem Approach of Aquatic Ecology, Mirror Lake and its Environment* (ed. G. E. Likens), Springer-Verlag, New York, pp. 345–66.

Davis, M. B., Moeller, R. E. and Ford, J. (1984) Sediment focusing and pollen influx, in *Lake Sediments and Environmental History* (eds E. Y. Haworth and J. W. G. Lund), University of Leicester Press, Leicester, pp. 261–93.

Davis, M. B., Moeller, R. E., Likens, G. E., Ford, M. S., Sherman, J. and Goulden, C. (1985b) Paleoecology of Mirror Lake and its watershed, in *An Ecosystem Approach of Aquatic Ecology, Mirror Lake and its Environment* (ed. G. E. Likens), Springer-Verlag, New York, pp. 410–29.

Davis, M. B., Spear, R. W. and Shane, L. C. K. (1980) Holocene climate of New England. *Quat. Res.*, **14**, 240–50.

Davis, M. B., Woods, K. D., Webb, S. L. and Futyma, R. P. (1986) Dispersal versus climate: expansion of *Fagus* and *Tsuga* into the Upper Great Lakes region. *Vegetatio*, **67**, 93–103.

Davis, O. K., Sheppard, J. C. and Robertson, S. (1986) Contrasting climatic histories for the Snake River Plain, Idaho, resulting from multiple thermal maxima. *Quat. Res.*, **26**, 321–39.

Davis, R. B. (1967) Pollen studies of near-surface sediments in Maine lakes, in *Quaternary Paleoecology* (eds E. J. Cushing and H. E. Wright, Jr.), Yale University Press, New Haven, pp. 143–73.

Davis, R. B. (1987) Paleolimnological diatom studies of acidification of lakes by acid rain: an application of Quaternary science. *Quat. Sci. Rev.*, **6**, 147–63.

Davis, R. B. and Jacobson, G. L. (1985) Late glacial and early Holocene landscapes in northern New England and adjacent areas of Canada. *Quat. Res.*, **23**, 341–68.

Deevey, E. S. (1942) Studies on Connecticut Lake sediments. III. The Biostratonomy of Linsley Pond. *Amer. J. Sci.*, **240**, 233–64, 313–24.

Deevey, E. S. (1969) Coaxing history to conduct experiments. *Bioscience*, **19**, 40–3.

Deevey, E. S. (1984) Stress, strain, and stability of lacustrine ecosystems, in *Lake Sediments and Environmental History* (eds E. Y. Haworth and J. W. G. Lund), University of Minnesota Press, Minneapolis, pp. 208–29.

Delcourt, H. R. (1979) Late-Quaternary vegetation history of the eastern Highland Rim and adjacent Cumberland Plateau of Tennessee. *Ecol. Monogr.*, **49**, 255–80.

Delcourt, H. R. (1987) The impact of prehistoric agriculture and land occupation on natural vegetation. *Trends Ecol. Evol.*, **2**, 39–44.

Delcourt, H. R. and Delcourt, P. A. (1985) Comparison of taxon calibrations, modern analogue techniques, and forest-stand simulation models for the quantitiative reconstrution of past vegetation. *Earth Surf. Proc. Landforms*, **10**, 293–304.

Delcourt, H. R. and Delcourt, P. A. (1986) Late-Quaternary vegetational history of the central Atlantic states, in *The Quaternary of Virginia* (eds J. McDonald and S. O. Bird), Virginia Commonwealth Division of Mineral Resources, Charlottesville, pp. 23–35.

Delcourt, H. R. and Delcourt, P. A. (1988) Quaternary landscape ecology: relevant scales in space and time. *Landscape Ecol.*, **2**, 23–44.

Delcourt, H. R., Delcourt, P. A. and Spiker, E. C. (1983) A 12 000-year record of forest history from Cahaba Pond, St. Clair County, Alabama. *Ecology*, **64**, 874–87.

Delcourt, H. R., Delcourt, P. A. and Webb, T. III. (1983) Dynamic plant ecology: the spectrum of vegetational change in space and time. *Quat. Sci. Rev.*, **1**, 153–175.

Delcourt, H. R., Delcourt, P. A., Wilkins, G. R., and Smith, E. N., Jr. (1986b) Vegetational history of the cedar glades regions of Tennessee, Kentucky, and Missouri during the past 30 000 years. *Assoc. Southeastern Biol. Bull.*, **33**, 128–37.

Delcourt, P. A. (1980) Goshen Springs: late-Quaternary vegetation record for southern Alabama. *Ecology*, **61**, 371–86.

Delcourt, P. A. and Delcourt, H. R. (1980) Pollen preservation and Quaternary environmental history in the southeastern United States. *Palynology*, **4**, 215–31.

Delcourt, P. A. and Delcourt, H. R. (1981) Vegetation maps for eastern North America: 40 000 yr B.P. to the present, in *Geobotany II* (ed. R. Romans), Plenum, New York, pp. 123–66.

Delcourt, P. A. and Delcourt, H. R. (1983) Late-Quaternary vegetational dynamics and community stability reconsidered. *Quat. Res.*, **19**, 265–71.

Delcourt, P. A. and Delcourt, H. R. (1984) Late-Quaternary paleoclimates and biotic responses across eastern North America and the northwestern Atlantic Ocean. *Palaeogeogr. Palaeoclimatol. Palaeoecol.*, **48**, 263–84.

Delcourt, P. A. and Delcourt, H. R. (1987a) *Long-term Forest Dynamics of the Temperate Zone, Ecological Studies 63*. Springer-Verlag, New York.

Delcourt, P. A. and Delcourt, H. R. (1987b) Late-Quaternary dynamics of temperate forests: applications of paleoecology to issues of global environment change. *Quat. Sci. Rev.*, **6**, 129–46.

Delcourt, P. A. and Delcourt, H. R. (1991) Ecotone dynamics in space and time, in *Landscape Boundaries: Consequences for Biotic Diversity and Ecological Flows* (eds F. di Castri and A. J. Hansen), Springer-Verlag, New York (in press).

Delcourt, P. A., Delcourt, H. R. and Webb, T. III. (1984) Atlas of mapped distributions of dominance and modern pollen percentages for important tree taxa of eastern North America. *Am. Assoc. Strat. Palynol. Contrib. Ser.*, **14**, 1–131.

Delcourt, P. A. Delcourt H. R., Brister, R. C. and Lackey, L. E. (1980) Quaternary vegetation history of the Mississippi Embayment. *Quat. Res.*, **13**, 111–32.

Delcourt, P. A., Delcourt, H. R., Cridlebaugh, P. A. and Chapman, J. (1986a) Holocene ethnobotanical and paleoecological record of human impact on vegetation in the Little Tennessee River Valley, Tennessee, U.S.A. *Quat. Res.*, **25**, 330–49.

Delmas, R. J., Ascencia, J. M. and Legrand, M. (1980) Polar ice evidence that atmospheric CO_2 20 000 B.P. was 50% of present. *Nature*, **284**, 155–7.

Diamond, J. (1986) Overview: laboratory experiments, field experiments, and natural experiments in *Community Ecology* (eds J. Diamond and T. J. Case), Harper and Row, New York, pp. 3–22.

Drury, W. H. and Nisbet, I. C. T. (1973) Succession. *J. Arnold Arboretum*, **54**, 331–68.

Dunwiddie, P. W. (1986) A 6000-year record of forest history on Mount Rainier, Washington. *Ecology*, **67**, 58–68.

Dunwiddie, P. W. (1987) Macrofossil and pollen representation of coniferous trees in modern sediments from Washington. *Ecology*, **68**, 1–11.

Dyer, M. I., di Castri, F. and Hansen, A. J. (1988) Geosphere–biosphere observatories: their definition and design for studying global change. *Biol. Internat.*, **16**, 1–40.

Elias, S. A. and Wilkinson, B. (1983) Lateglacial insect fossil assemblages from Lobsigensee (Swiss Plateau). *Rev. Paleobiol.*, **2**, 189–204.

Elton, C. S. (1958) *The Ecology of Invasions*, Chapman and Hall, London. (Reprinted in 1972.)

Emanuel, W. R., Shugart, H. H. and Stevenson, M. P. (1985a) Climatic change and the broad-scale distribution of terrestrial ecosystem complexes. *Climatic Change*, **7**, 29–43.

Emanuel, W. R., Shugart, H. H. and Stevenson, M. P. (1985b) Response to comment: climatic change and the broad-scale distribution of terrestrial ecosystem complexes. *Climatic Change*, **7**, 457–60.

Emlen, J. M. (1973) *Ecology: an Evolutionary Approach*, Addison-Wesley, Reading, MA.

Faegri, K. and Iversen, J. (1975) *Textbook of Pollen Analysis*, 3rd edn. Hafner, New York.

Ford, M. S. (J.) (1990) A 10 000-yr history of natural ecosystem acidification. *Ecol. Monogr.*, **60**, 57–89.

Foster, D. R. and Wright, H. E., Jr (1990) Role of ecosystem development and climate change in bog formation in central Sweden. *Ecology*, **7**, 450–63.

Foster, D. R., Wright, H. E., Jr., Thelaus, M. and King, G. A. (1988) Bog development and landform dynamics in central Sweden and south-eastern Labrador, Canada. *J. Ecol.*, **76**, 1164–85.

Fowells, H. A. (1965) Silvics of forest trees of the United States. *U. S. Forest Service Agric. Handbk.*, **271**, 1–762.

Frissell, S. S. (1973) The importance of fire as a natural ecological factor in Itasca State Park, Minnesota. *Quat. Res.*, **3**, 397–407.

Gaillard, M.-J. (1983) On the occurrence of *Betula nana* L. pollen grains in the Late-Glacial deposits of Lobsigensee (Swiss Plateau). *Rev. Paleobiol*, **2**, 163–80.

Gajewski, K. (1987) Climatic impacts on the vegetation of eastern North America during the past 2000 years. *Vegetatio*, **68**, 179–90.

Gajewski, K., Swain, A. M. and Peterson, G. M. (1987) Late Holocene pollen stratigraphy in four northeastern United States lakes. *Geogr. Phys. Quat.*, **41**, 377–86.

Gajewski, K., Winkler, M. G. and Swain, A. M. (1985) Vegetation and fire history from three lakes with varved sediments in northwestern Wisconsin (USA). *Rev. Palaeobot. Palynol.*, **44**, 277–92.

Gammon, R. H., Sundquist, E. T. and Fraser, P. J. (1985) History of carbon dioxide in the atmosphere, in *Atmospheric Carbon Dioxide and the Global Carbon Cycle* (ed. J. R. Trabalka), U.S. Department of Energy, Washington, DC, pp. 25–62.

Gauch, H. (1982) *Multivariate Analysis in Community Ecology*, Cambridge University Press, Cambridge.

Gensel, P. G. and Andrews, H. N. (1987) The evolution of early land plants. *Am. Scientist*, **75**, 478–89.

Genthon, C., Barnola, J. M., Raynaud, D., Lorius, C., Jouzel, J., Barkov, N. I., Korotkevich, Y. S. and Kotlyakov, V. M. (1987) Vostok ice core: climatic response to CO_2 and orbital forcing changes over the last climatic cycle. *Nature*, **329**, 414–8.

Glaessner, M. F. (1984) *The Dawn of Animal Life: a Biohistorical Study*, Cambridge University Press, Cambridge.

Glaser, P. H. (1981) Transport and deposition of leaves and seeds on tundra: a late-glacial analog. *Arctic Alpine Res.*, **13**, 173–82.

Gleason, H. A. (1926) The individualistic concept of plant association. *Bull. Torrey Bot. Club*, **53**, 7–26.

Goudie, A. (1981) *Geomorphological Techniques*, George Allen and Unwin, London.

Gould, S. J. (1965) Is uniformitarianism necessary? *Amer. J. Sci.*, **263**, 223–8.

Goulden, C. and Vostreys, G. (1985) Paleolimnology: animal microfossils, in *An Ecosystem Approach of Aquatic Ecology, Mirror Lake and its Environment* (ed. G. E. Likens), Springer-Verlag, New York, pp. 382–6.

Graham, R. W. (1976) Late Wisconsin mammalian faunas and environmental gradients of the eastern United States. *Paleobiology*, **2**, 343–50.

Graham, R. W. (1986) Response of mammalian communities to environmental changes during the Late Quaternary, in *Community Ecology* (eds J. Diamond and T. J. Case), Harper and Row, New York, pp. 300–13.

Graham, R. W. and Lundelius, E. L., Jr. (1984) Coevolutionary disequilibrium and Pleistocene extinctions, in *Quaternary Extinctions – a Prehistoric Revolution* (eds. P. S. Martin and R. G. Klein), University of Arizona Press, Tucson, pp. 223–49.

Graham, R. W., Semken, H. A., Jr. and Graham, M. A. (editors) (1987) *Late Quaternary Mammalian Biogeography and Environments of the Great Plains and Prairies*, Illinois State Museum, Springfield, IL.

Grayson, D. K. (1984) *Quantitative Zooarchaeology: Topics in the Analysis of Archaeological Fauna*, Academic Press, Orlando.

Green, D. G. (1981) Times series and postglacial forest ecology. *Quat. Res.*, **15**, 265–77.

Green, D. G. (1982) Fire and stability in the postglacial forests of southwest Nova Scotia. *J. Biogeogr.*, **9**, 29–40.

Grime, J. P. (1979) *Plant Strategies and Vegetation Processes*, John Wiley and Sons, Chichester.

Grimm, E. C. (1983) Chronology and dynamics of vegetation change in the prairie–woodland region of southern Minnesota, USA. *New Phytol.*, **93**, 311–50.

Grimm, E. C. (1984) Fire and other factors controlling the vegetation of the Big Woods region of Minnesota. *Ecol. Monogr.*, **54**, 291–311.

Guries, R. P. and Nordheim, E. V. (1984) Flight characteristics and dispersal potential of maple samaras. *For. Sci.*, **30**, 434–40.

Guthrie, R. D. (1968) Paleoecology of the large-mammal community in interior Alaska during the late Pleistocene. *Amer. Midl. Nat.*, **79**, 346–63.

Guthrie, R. D. (1982) Mammals of the mammoth steppe as paleoenvironmental indictors, in *Paleoecology of Beringia* (eds D. M. Hopkins, J. V. Matthews, Jr., C. E. Schweger, and S. B. Young), Academic Press, London, pp. 307–26.

Guthrie, R. D. (1984) Mosaics, allelochemics and nutrients, an ecological theory of late Pleistocene megafaunal extinctions, in *Quaternary Extinctions* (eds. P. S. Martin and R. G. Klein), University of Arizona Press, Tucson, pp. 259–98.

Hall, S. A. (1981) Deteriorated pollen grains and the interpretation of Quaternary pollen diagrams. *Rev. Palaeobot. Palynol.*, **32**, 193–206.

Hansel, A. K. and Mickelson, D. M. (1988) A reevaluation of the timing and causes of high lake phases in the Lake Michigan Basin. *Quat. Res.*, **29**, 113–28.

Hansel, A. K., Mickelson, D. M., Schneider, A. F. and Larsen, C. E. (1985) Late Wisconsin and Holocene history of the Lake Michigan Basin, in *Quaternary Evolution of the Great Lakes* (eds P. F. Karrow and P. E. Calkin), Geological Association of Canada Special Paper 30, pp. 39–53.

Hansen, J., Russell, G., Rind, D., Stone, P., Lacis, A., Lebedeff, S., Ruedy, R. and Travis, L. (1983) Efficient 3-dimensional global models for climate studies: models I and II. *April Monthly Weather Rev.*, **3**, 609–62.

Harper, J. L. (1977) *Population Biology of Plants*, Academic Press, London.

Haworth, E. Y. and Lund, J. W. G. (eds) (1984) *Lake Sediments and Environmental History*, University of Minnesota Press, Minneapolis.

Healy, P. F., Lambert, J. D. H., Arnason, J. T. and Hebda, R. J. (1983) Caracol, Belize: evidence of ancient Maya agricultural terraces. *J. Field Archaeol.*, **10**, 397–410.

Heinselman, M. L. (1973) Fire in the virgin forests of the Boundary Waters Canoe Area, Minnesota. *Quat. Res.*, **3**, 329–82.

Hofmann, W. (1983) Stratigraphy of subfossil Chironomidae and Ceratopogonidae (Insecta: Diptera) in Late-Glacial sediments from Lobsigensee (Swiss Plateau). *Rev. Paleobiol.*, **2**, 205–9.

Hopkins, D. M., Matthews, J. V., Jr., Schweger, C. E. and Young, S. B. (1982) *Paleoecology of Beringia*, Academic Press, London.

Horn, H. S. (1976) Succession, in *Theoretical Ecology: Principles and Applications* (ed. R. M. May), Belknap Press, Cambridge, MA, pp. 196–211.

Hubbell, S. P. and Foster, R. B. (1986) Biology, chance, and history and the structure of tropical rain forest tree communities, Chapter 19, in *Community Ecology* (eds J. Diamond and T. J. Case), Harper and Row, New York, pp. 314–29.

Huntley, B. (1988) Europe, in *Vegetation History* (eds B. Huntley and T. Webb, III), Kluwer, Dordrecht, pp. 341–83.

Huntley, B. and Birks, H. J. B. (1983) *An Atlas of Past and Present Pollen Maps for Europe: 0–13 000 Years Ago*. Cambridge University Press, Cambridge.

Huntley, B. and Webb, T. III (editors) (1988) *Vegetation History*, Kluwer, Dordrecht.

Huntley, B., Bartlein, P. J. and Prentice, I. C. (1989) Climatic control of the distribution and abundance of beech (*Fagus* L.) in Europe and North America. *J. Biogeogr.*, **16**, 551–60.

Hurlbert, S. H. (1984) Pseudoreplication and the design of ecological field experiments. *Ecol. Monogr.*, **54**, 187–211.

Hustedt, F. (1939) Systematische und Okologische Untersuchungen uber die Diatomeenflora von Java, Bali, und Sumatra nach dem Material der Dekutschen Limnologischen Sunda-Expedition III. Die Okologischen Factoren und ihr Einfluss auf die Diatomeenflora. *Arch. Hydrobiol.* (Supplement), **16**, 274–394.

Hutson, W. H. (1977) Transfer functions under no-analog conditions: experiments with Indian Ocean planktonic Foraminifera. *Quat. Res.*, **8**, 355–67.

Hyvärinen, H. (1975) Absolute and relative pollen diagrams from northwest Fenno-scandia. *Fennia*, **142**, 1–23.

Imbrie, J. and Imbrie, K. P. (1979) *Ice Ages: Solving the Mystery*, Enslow Publishers, Short Hills, NJ.

Iversen, J. (1954) The late-glacial flora of Denmark and its relation to climate and soil. *Danmarks Geologiske Undersogelse*, II, **80**, 87–119.

Iversen, J. (1958) The bearing of glacial and interglacial epochs on the formation and extinction of plant taxa. *Uppsala Universiteit Arssk*, **6**, 210–5.

Iversen, J. (1969) Retrogressive development of a forest ecosystem demonstrated by pollen diagrams from fossil mor. *Oikos* (Suppl.), **12**, 35–49.

Jackson, S. T., Futyma, R. P. and Wilcox, D. A. (1988) A paleoecological test of a classical hydrosere in the Lake Michigan dunes. *Ecology*, **69**, 928–36.

Jacobson, G. L. (1979) The palaeoecology of white pine (*Pinus strobus*) in Minnesota. *J. Ecol.*, **67**, 697–726.

Jacobson, G. L. and Bradshaw, R. H. W. (1981) The selection of sites for paleovegetational studies. *Quat. Res.*, **16**, 80–96.

Jacobson, G. L., and Grimm, E. C. (1986) A numerical analysis of Holocene forest and prairie vegetation in central Minnesota. *Ecology*, **67**, 958–66.

Jacobson, G. L., Webb, T., III and Grimm, E. C. (1987) Patterns and rates of vegetation change during the deglaciation of eastern North America, in *North America and Adjacent Oceans During the Last Deglaciation*, Vol. K-3, *The Geology of North America* (eds W. F. Ruddiman and H. E. Wright, Jr.), Geological Society of America, Boulder, pp. 277–88.

Janssen, C. R. (1966) Recent pollen spectra from the deciduous and coniferous–deciduous forests of northwestern Minnesota, interpreted from pollen indicators and surface samples. *Ecol. Monogr.*, **37**, 145–72.

Janzen, D. H. (1970) Herbivores and the number of tree species in tropical forests. *Amer. Nat.*, **104**, 501–28.

Janzen, D. H. (1983) *Costa Rican Natural History*, University of Chicago Press, Chicago.

Janzen, D. H. and Martin, P. S. (1982) Neotropical anachronisms: the fruits the Gomphotheres ate. *Science*, **215**, 19–27.

Jenny, H., Arkley, R. J. and Schultz, A. M. (1969) The pygmy forest-podzol ecosystem and its dune associates of the Mendocino Coast. *Madrono*, **20**, 60–74.

Johnson, W. C. and Adkisson, C. S. (1985) Dispersal of beech nuts by blue jays in fragmented landscapes. *Amer. Midl. Nat.*, **113**, 319–24.

Johnson, W. C. and Adkisson, C. S. (1986) Airlifting the Oaks. *Nat. Hist.*, **10**, 41–6.

Johnson, W. C. and Webb, T., III (1989) The role of blue jays (*Cyanocitta cristata* L.) in the postglacial dispersal of fagaceous trees in eastern North America. *J. Biogeogr.*, **16**, 561–71.

Johnson, W. C., Sharpe, D. M., DeAngelis, D. L., Fields, D. E. and Olson, R. J. (1981) Modeling seed dispersal and forest island dynamics, in *Forest Island Dynamics in Man-dominated Landscapes* (eds R. L. Burgess and D. M. Sharpe), Springer-Verlag, New York, pp. 215–39.

Kapp, R. O., Bushouse, S. and Foster, B. (1969) A contribution to the geology and forest history of Beaver Island, Michigan. Proceedings of the 12th Conference on Great Lakes Research, 225–36.

Kapp, R. O. (1977) Late Pleistocene and postglacial plant communities of the Great Lakes region, in *Geobotany* (ed. R. Romans), Plenum, New York, pp. 1–27.

Keever, C. (1953) Present composition of some stands of the former oak-chestnut forests in the southern Blue Ridge Mountains. *Ecology*, **34**, 44–5.

Kellogg, W. W. and Schware, R. (1981) *Climate Change and Society: Consequences of Increasing Atmospheric Carbon Dioxide*, Westview Press, Boulder, CO.

Kellogg, W. W. and Schware, R. (1982) Society, science, and climate change. *Foreign Affairs*, **60**, 1076–109.

Kershaw, A. P. (1981) Quaternary vegetation and environments, in *Ecological Biogeography of Australia*, Vol. 1 (ed. A. Keast), Dr W. Junk, Dordrecht, pp. 81–101.

King, F. B. (1985) Early cultivated Cucurbits in eastern North America, in *Prehistoric Food Production in North America, Anthropological Papers No. 75* (ed. R. I. Ford), Museum of Anthropology, University of Michigan, Ann Arbor, pp. 73–97.

King, J. E. (1981) Late Quaternary vegetational history of Illinois. *Ecol. Monogr.*, **51**, 43–62.

Kurten, B. (1986) *How to Deep-Freeze a Mammoth*, Columbia University Press, New York.

Kurten, B. and Anderson, E. (1980) *Pleistocene Mammals of North America*, Columbia University Press, New York.

Kutzbach, J. E. and Guetter, P. J. (1986) The influence of changing orbital parameters and surface boundary conditions on climate simulations for the past 18 000 years. *J. Atmosph. Sci.*, **43**, 1726–59.

Kutzbach, J. E. and Wright, H. E., Jr. (1985) Simulation of the climate of 18 000 years BP: results for the North American/North Atlantic/European sector and comparison with the geologic record of North America. *Quat. Sci. Rev.*, **4**, 147–87.

Lachenbruch, A. H. and Marshall, B. V. (1986) Changing climate: geothermal evidence from permafrost in the Alaskan arctic. *Science*, **234**, 689–96.

Leyden, B. W. (1984) Guatemalan forest synthesis after Pleistocene aridity. *Proc. Nat. Acad. Sci. USA*, **81**, 4856–9.

Libby, W. F. (1955) *Radiocarbon Dating*, University of Chicago Press, Chicago.

Likens, G. E. (1985) Mirror Lake: cultural history, in *An Ecosystem Approach of Aquatic Ecology, Mirror Lake and its Environment* (ed. G. E. Likens), Springer-Verlag, New York, pp. 72–83.

Likens, G. E. and Bormann, F. H. (1974) Linkages between terrestrial and aquatic ecosystems. *Bioscience*, **24**, 447–56.

Likens, G. E. and Davis, M. B. (1975) Post-glacial history of Mirror Lake and its watershed in New Hampshire, USA: an initial report. *Verh. Internat. Verein. Limnol.*, **19**, 982–93.

Likens, G. E. and Moeller, R. E. (1985a) Paleolimnology: fossil pigments, in *An Ecosystem Approach of Aquatic Ecology, Mirror Lake and its Environment* (ed. G. E. Likens), Springer-Verlag, New York, pp. 387–91.

Likens, G. E. and Moeller, R. E. (1985b) Paleolimnology: chemistry, in *An Ecosystem Approach of Aquatic Ecology, Mirror Lake and its Environment* (ed. G. E. Likens), Springer-Verlag, New York, pp. 392–410.

Likens, G. E., Bormann, F. H., Pierce, R. S., Eaton, J. S. and Johnson, N. M. (1977) *Biogeochemistry of a Forested Ecosystem*, Springer-Verlag, New York.

Lindeman, R. L. (1941) The developmental history of Cedar Bog Lake. *Amer. Midl. Nat.*, **26**, 101–12.

Lindeman, R. L. (1942) The trophic-dynamic aspect of ecology. *Ecology*, **23**, 399–418.

Little, E. L., Jr. (1971) *Atlas of United States Trees*, Vol. 1, *Conifers and Important Hardwoods*, US Dept. Agric. For. Serv. Misc. Publ. 1146 (8 p. and 200 maps).

Lomolino, M. V., Brown, J. H. and Davis, R. (1989) Island biogeography of montane forest mammals in the American Southwest. *Ecology*, **70**, 180–94.

Loucks, O. L. (1970) Evolution of diversity, efficiency, and community stability. *Amer. Zool.*, **10**, 17–25.

Lundelius, E. L., Jr., Graham, R. W., Anderson, E., Guilday, J., Holman, J. A., Steadman, D. W. and Webb, S. D. (1983) Terrestrial vertebrate faunas, in *Late-Quaternary Environments of the United States*, Vol. 1, *The Late Pleistocene* (ed. S. C. Porter), University of Minnesota Press, Minneapolis, pp. 311–53.

MacArthur, R. H. and Wilson, E. O. (1967) The theory of island biogeography. *Princeton Monogr. Pop. Biol.*, **1**, 1–203.

MacDonald, G. M. (1987) Postglacial development of the subalpine-boreal transition forest of western Canada. *J. Ecol.*, **75**, 303–20.

MacDonald, G. M. and Cwynar, L. C. (1985) A fossil pollen based reconstruction of the late Quaternary history of lodgepole pine (*Pinus contorta* ssp. *latifolia*) in the Western interior of Canada. *Can. J. For. Res.*, **15**, 1039–44.

MacDonald, G. M. and Ritchie, J. C. (1986) Modern pollen surface samples and the interpretation of postglacial vegetation development in the western interior of Canada. *New Phytol.*, **103**, 245–68.

Mack, R. N. (1981) Invasion of *Bromus tectorum* L. into western North America: an ecological chronicle. *Agro-Ecosystems*, **7**, 145–65.

Mack, R. N. (1985) Invading plants: their potential contribution to population biology, in *Studies on Plant Demography* (ed. J. White), Academic Press, London, pp. 127–42.

Mahaney, W. C. (1984) *Quaternary Dating Methods*, Elsevier, Amsterdam.

Manabe, S. and Wetherald, R. T. (1987) Large-scale changes in soil wetness induced by an increase in carbon dioxide. *J. Atmosph. Sci.*, **44**, 1211–35.

Markgraf, V. (1986) Plant inertia reassessed. *Amer. Nat.*, **127**, 725–6.

Marks, P. L. (1983) On the origin of the field plants of the northeastern United States. *Amer. Nat.*, **122**, 210–28.

Marschner, F. J. (1974) Map of the original vegetation of Minnesota, Scale 1:500 000 (redrafted by P. J. Burwell and S. J. Haas, with accompanying text by M. L. Heinselman). North Central For. Expt. Sta., St. Paul, Minnesota.

Martin, P. S. and Klein, R. G. (eds) (1984) *Quaternary Extinctions: a Prehistoric Revolution*, University of Arizona Press, Tucson.

Matthews, J. V., Jr. (1982) East Beringia during Late Wisconsin time: a review of the biotic evidence, in *Paleoecology of Beringia* (eds D. M. Hopkins, J. V. Matthews, Jr., C. E. Schweger and S. B. Young), Academic Press, London, pp. 127–52.

McAndrews, J. H. (1966) Postglacial history of prairie, savanna, and forest in northwestern Minnesota. *Mem. Torrey Bot. Club*, **22**, 1–72.

McAndrews, J. H. (1968) Pollen evidence for the protohistoric development of the 'Big Woods' in Minnesota (USA). *Rev. Palaeobot. Palynol.*, **7**, 201–11.

McAndrews, J. H. (1988) Human disturbance of North American forests and grasslands: the fossil pollen record, in *Vegetation History* (eds B. Huntley and T. Webb, III), Kluwer, Dordrecht, pp. 673–97.

McCormick, J. F. and Platt, R. B. (1980) Recovery of an Appalachian forest following the chestnut blight or Catherine Keever – you were right! *Amer. Midl. Nat.*, **104**, 264–73.

McIntosh, R. P. (1985) *The Background of Ecology: Concept and Theory*, Cambridge University Press, Cambridge.

Mehringer, P. J., Jr., Blinman, E. and Petersen, K. L. (1977) Pollen influx and volcanic ash. *Science*, **198**, 257–61.

Merilainen, J., Huttunen, P. and Battarbee, R. W. (eds) (1983) *Paleolimnology*. Dr W. Junk, Dordrecht.

Miller, N. G. (1973) Lateglacial plants and plant communities in northwestern New York State. *J. Arnold Arb.*, **54**, 123–59.

Miller, N. G. (1980) Mosses as paleoecological indicators of lateglacial terrestrial environments: some North American studies. *Bull. Torrey Bot. Club*, **107**, 373–91.

Mitchell, J. F. B. (1983) The seasonal response of a general circulation model to changes in CO_2 and sea temperatures. *Quart. J. Royal Meteorol. Soc.*, **109**, 113–52.

Mitchell, J. F. B. and Lupton, G. (1984) A $4 \times CO_2$ integration with prescribed changes in sea surface temperatures. *Prog. Biometeorol.*, **3**, 353–74.

Moeller, R. E. (1985) Macrophytes, in *An Ecosystem Approach of Aquatic Ecology, Mirror lake and its Environment* (ed. G. E. Likens), Springer-Verlag, New York, pp. 177–92.

Mooney, H. A. (1988) Ecologists and the Global Change Program. *Trends Ecol. Evol.*, **3**, 4–5.

Mooney, H. A. and Drake, J. A. (1986) *Ecology of Biological Invasions of North America and Hawaii, Ecological Studies 58*, Springer-Verlag, New York.

Moore, P. D. (1975) Origin of blanket mires. *Nature*, **256**, 267–9.

Moore, P. D. (ed.) (1984) *European Mires*, Academic Press, London.

Moran, V. C. and Annecke, D. P. (1979) Critical reviews of biological pest control in South Africa. 3. The jointed cactus, *Opuntia aurantiaca* Lindley. *J. Entomol. Soc. South Africa*, **42**, 299–329.

National Reseach Council (1986) *Global Change in the Geosphere–Biosphere: Initial Priorities for an IGBP*, National Academy Press, Washington, DC.

National Research Council (1988) *Toward an Understanding of Global Change: Initial Priorities for U.S. Contributions to the International Geosphere–Biosphere Program*, National Academy Press, Washington, DC.

Neftel, A., Oeschger, H., Schwander, J., Stauffer, B. and Zumbrunn, R. (1982) Ice core measurements give atmospheric CO_2 content during the past 40 000 y. *Nature*, **295**, 220–3.

Neftel, A., Moor, E., Oeschger, H. and Stauffer, B. (1985) Evidence from polar ice cores for the increase in atmospheric CO_2 in the last two centuries. *Nature*, **315**, 45–7.

Nei, M., Maruyama, T. and Chakraborty, R. (1975) The bottleneck effect and genetic variability in populations. *Evolution*, **29**, 1–10.

Nitecki, M. H. (editor) (1981) *Biotic Crises in Ecological and Evolutionary Time*, Academic Press, New York.

Odum, E. P. (1969) The strategy of ecosystem development. *Science*, **164**, 262–70.

Oliver, C. D. (1981) Forest development in North America following major disturbances. *For. Ecol. Manage.*, **3**, 153–68.

Olson, J. S. (1958) Rates of succession and soil changes on southern Lake Michigan sand dunes. *Bot. Gaz.*, **119**, 125–70.

O'Neill, R. V., DeAngelis, D. L., Waide, J. B. and Allen, T. F. H. (1986) A hierarchical concept of ecosystems. *Princeton Monogr. Pop. Biol.*, **23**, 1–253.

Osvald, H. (1923) Die Vegetation der Ozeanischen Hochmoore in Norwegen. *Svenska Vaxsociologiska Sallskapets Handlingar*, **VII**, 106 p.

Overpeck, J. T., Rind, D. and Goldberg, R. (1990) Climate-induced changes in forest disturbance and vegetation. *Nature*, **343**, 51–3.

Overpeck, J. T., Webb, T. III and Prentice, I. C. (1985) Quantitative interpretation of fossil pollen spectra: dissimilarity coefficients and the method of modern analogs. *Quat. Res.*, **23**, 87–108.

Patterson, W. A. III and Backman, A. E. (1988) Fire and disease history of forests, in *Vegetation History* (eds B. Huntley and T. Webb, III), Kluwer, Dordrecht, pp. 603–32.

Patterson, W. A. III, Edwards, K. J. and Maguire, D. J. (1987) Microscopic charcoal as a fossil indicator of fire. *Quat. Sci. Rev.*, **6**, 3–23.

Patrick, R. (1943) The diatoms of Linsley Pond, Connecticut. *Proc. Acad. Nat. Sci. Philadelphia*, **95**, 53–110.

Payette, S. (1988) Late-Holocene development of subarctic ombrotrophic peatlands: allogenic and autogenic succession. *Ecology*, **69**, 516–31.

Payette, S. and Gagnon, R. (1985) Late Holocene deforestation and tree regeneration in the forest–tundra of Quebec. *Nature*, **313**, 570–2.

Payette, S., Morneau, C., Sirois, L. and Desponts, M. (1989) Recent fire history of the northern Quebec biomes. *Ecology*, **70**, 656–73.

Peet, R. K. and Christensen, N. L. (1980) Succession: a population process. *Vegetatio*, **43**, 131–40.

Peters, R. L. and Darling, J. D. S. (1985) The greenhouse effect and nature reserves. *Bioscience*, **35**, 707–17.

Pickett, S. T. A. (1976) Succession: an evolutionary interpretation. *Amer. Nat.*, **110**, 107–19.

Pickett, S. T. A. (1989) Space-for-time substitution as an alternative to long-term studies, in *Long-Term Studies in Ecology, Approaches and Alternatives* (ed. G. E. Likens), Springer-Verlag, New York, pp. 110–35.

Pickett, S. T. A. and White, P. S. (eds) (1985) *The Ecology of Natural Disturbance and Patch Dynamics*, Academic Press, New York.

Pielou, E. C. (1979) *Biogeography*, John Wiley and Sons, New York.

Pimm, S. L. (1982) *Food Webs*, Chapman and Hall, London.

Pimm, S. L. (1984) The complexity and stability of ecosystems. *Nature*, **307**, 321–6.

Pisias, N. G. and Moore, T. C., Jr. (1981) The evolution of Pleistocene climate: a time series approach. *Earth Planet Sci. Let.*, **52**, 450–8.

Porter, S. C. (ed.) (1983) *Late-Quaternary Environments of The United States*, Vol. 1, *The Late Pleistocene*, University of Minnesota Press, Minneapolis.

Prentice, I. C. (1985) Pollen representation, source area, and basin size: toward a unified theory of pollen analysis. *Quat. Res.*, **23**, 76–86.

Prentice, I. C. (1986a) Multivariate methods for data analysis, in *Handbook of Holocene Palaeoecology and Palaeohydrology* (ed. B. E. Berglund), Wiley and Sons, New York, pp. 775–97.

Prentice, I. C. (1986b) Vegetation responses to past climatic variation. *Vegetatio*, **67**, 131–41.

Prentice, I. C. (1988) Records of vegetation in time and space: the principles of pollen analysis, in *Vegetation History* (eds B. Huntley and T. Webb, III), Kluwer, Dordrecht, pp. 17–42.

Redfield, A. C. (1965) Ontogeny of a salt marsh estuary. *Science*, **147**, 50–5.

Redfield, A. C. (1972) Development of a New England salt marsh. *Ecol. Monogr.*, **42**, 201–37.

Ritchie, J. C. (1985) Late-Quaternary climatic and vegetational change in the Lower Mackenzie Basin, northwest Canada. *Ecology*, **66**, 612–21.

Ritchie, J. C. (1986) Climate change and vegetation response. *Vegetatio*, **67**, 65–74.

Ritchie, J. C. (1987) *Postglacial Vegetation of Canada*, Cambridge University Press, Cambridge.

Ritchie, J. C. and Cwynar, L. C. (1982) The Late Quaternary vegetation of the North Yukon, in *Paleoecology of Beringia* (eds D. M. Hopkins, J. V. Matthews, Jr., C. E. Schweger, and S. B. Young), Academic Press, London, pp. 113–26.

Ritchie, J. C. and MacDonald, G. M. (1986) The patterns of post-glacial spread of white spruce. *J. Biogeogr.*, **13**, 527–40.

Roberts, L. (1989) How fast can trees migrate? *Science*, **243**, 735–7.

Robinson, T. W. (1965) Introduction, spread and areal extent of saltcedar (*Tamarix*) in the western states. *U.S. Geol. Surv. Prof. Paper* 491-A.

Romme, W. H. and Knight, D. H. (1982) Landscape diversity: the concept applied to Yellowstone Park. *Bioscience*, **32**, 644–70.

Ruddiman, W. F. and Wright, H. E., Jr. (editors) (1987) *North America and Adjacent Oceans During the Last Deglaciation*, Vol. K-3, *The Geology of North America*, Geological Society of America, Boulder, CO.

Sangster, A. G. and Dale, H. M. (1964) Pollen grain preservation of under-represented species in fossil spectra. *Can. J. Bot.*, **42**, 437–49.

Sauer, J. D. (1988) *Plant Migration: the Dynamics of Geographic Patterning in Seed Plant Species*, University of California Press, Berkeley and Los Angeles.

Schneider, S. H. (1989) The Greenhouse Effect: science and policy. *Science*, **243**, 771–81.

Schonewald-Cox, C. M. (1988) Boundaries in the protection of nature reserves. *Bioscience*, **38**, 480–9.

Sears, P. B. (1942) Postglacial migration of five forest genera. *Amer. J. Bot.*, **29**, 684–91.

Semken, H. A., Jr. (1983) Holocene mammalian biogeography and climatic change in the eastern and central United States, in *Late-Quaternary Environments of the United States*, Vol. 2, *The Holocene* (ed. H. E. Wright, Jr.), University of Minnesota Press, Minneapolis, pp. 182–207.

Shackleton, N. J. and Pisias, N. G. (1985) Atmospheric CO_2 orbital forcing and climate, in *The Carbon Cycle and Atmospheric CO_2: Natural Variations Archean to Present* (American Geophysical Union Monograph 32), (eds E. T. Sundquist and W. S. Broecker), American Geophysical Union, Washington, DC, pp. 303–18.

Shackleton, N. J., Hall, M. A., Line, J. and Chuxi, C. (1983) Carbon isotope data in Core V9-30 confirm reduced carbon dioxide concentration in Ice Age atmosphere. *Nature*, **306**, 319–22.

Shelford, V. E. (1911) Ecological succession. II. Pond fishes. *Biol. Bull.*, **21**, 127–51.

Shelford, V. E. (1913) *Animal Communities in Temperate America as Illustrated in the Chicago Region*, University of Chicago Press, Chicago.

Sherman, J. W. (1985) Paleolimnology: diatoms, in *An Ecosystem Approach of Aquatic Ecology, Mirror Lake and its Environment* (ed. G. E. Likens), Springer-Verlag, New York, pp. 366–82.

Shugart, H. H. (1984) *A Theory of Forest Dynamics*, Springer-Verlag, New York.

Shugart, H. H. and West, D. C. (1977) Development of an Appalachian deciduous forest succession model and its application to assessment of the impact of the chestnut blight. *J. Env. Manage.*, **5**, 161–79.

Smith, A. G. (1965) Problems of inertia and threshold related to post-Glacial habitat changes. *Proc. Roy. Soc. London*, **161**, 331–42.

Smol, J. P., Batterbee, R. W., Davis, R. B. and Merilainen, J. (eds) (1986) *Diatoms and Lake Acidity*, Dr W. Junk, Dordrecht.

Solomon, A. M. and Shugart, H. H. (1984) Integrating forest-stand simulations with paleoecological records to examine long-term forest dynamics, in *State and Change of Forest Ecosystems – Indicators in Current Research (Swed. Univ. Agric. Sci. Dept. Ecology and Environmental Research Report Nr. 13)* (ed. G. I. Agren), Uppsala, pp. 333–56.

Solomon, A. M. and Tharp, M. L. (1985) Simulation experiments with late Quaternary carbon storage in mid-latitude forests, in *The Carbon Cycle and*

Atmospheric CO₂: Natural Variations Archean to Present (eds E. T. Sundquist and W. S. Broecker), American Geophysical Union, Washington, DC, pp. 235–50.

Solomon, A. M. and Webb, T., III (1985) Computer-aided reconstruction of late-Quaternary landscape dynamics. *Ann. Rev. Ecol. Syst.*, **16**, 63–84.

Solomon, A. M. and West, D. C. (1985) Potential responses of forests to CO₂-induced climate change, in *Characterization of Information Requirements for Studies of CO₂ Effects: Water Resources, Agriculture, Fisheries, Forests and Human Health* (ed. M. R. White), U.S. Department of Energy, Washington, D.C., pp. 145–70.

Solomon, A. M. and West, D. C. (1986) Atmospheric carbon dioxide change: agent of future forest growth or decline, in *Effects of Changes in Stratospheric Ozone and Global Climate*, Vol. 3, *Climate Change* (ed. J. G. Titus), US Environmental Protection Agency, Washington, DC, pp. 23–38.

Solomon, A. M., West, D. C. and Solomon, J. A. (1981) Simulating the role of climate change and species immigration in forest succession, in *Forest Succession, Concepts and Application* (eds D. C. West, H. H. Shugart and D. B. Botkin), Springer-Verlag, New York, pp. 154–77.

Solomon, A. M., Delcourt, H. R., West, D. C. and Blasing, T. J. (1980) Testing a simulation model for reconstruction of prehistoric forest-stand dynamics. *Quat. Res.*, **14**, 275–93.

Spackman, W., Dolsen, C. P. and Riegel, W. (1966) Phytogenic organic sediments and sedimentary environments in the Everglades-mangrove complex. Part I: evidence of a transgressing sea and its effects on environments of the Shark River area of southwestern Florida. *Sonderdruck aus Palaeontogr. Beitr. zur Natur. der Vorzeit*, **117(B)**, 135–52.

Spear, R. W. (1989) Late-Quaternary history of high-elevation vegetation in the White Mountains of New Hampshire. *Ecol. Monogr.*, **59**, 125–51.

Stanley, S. M. (1979) *Macroevolution: Pattern and Process*, W. J. Freeman, San Francisco.

Stockmarr, J. (1975) Retrogressive forest development, as reflected in a mor pollen diagram from Mantingerbos, Brenthe, The Netherlands. *Palaeohistoria*, **17**, 38–51.

Stuckey, R. L. (1981) Origin and development of the concept of the Prairie Peninsula. *Ohio Biol. Surv. Biol. Notes*, **15**, 4–23.

Stuiver, M., Burk, R. L. and Quay, P. D. (1984) $^{13}C/^{12}C$ ratios and the transfer of biospheric carbon to the atmosphere. *J. Geophys. Res.*, **89**, 11731–48.

Swain, A. M. (1973) A history of fire and vegetation in northeast Minnesota as recorded in lake sediments. *Quat. Res.*, **3**, 383–96.

Swain, A. M. (1978) Environmental changes during the past 2000 years in north-central Wisconsin: analysis of pollen, charcoal, and seeds from varved lake sediments. *Quat. Res.*, **10**, 55–68.

Swank, W. T. and Crossley, D. A., Jr. (editors) (1988) *Forest Hydrology and Ecology at Coweeta, Ecological Studies 66*, Springer-Verlag, New York.

Swanson, F. J., Fratz, T. K., Caine, N. and Woodmansee, R. G. (1988) Landform effects on ecosystem patterns and processes. *Bioscience*, **38**, 92–8.

Tansley, A. G. (1935) The use and abuse of vegetational terms and concepts. *Ecology*, **16**, 284–307.

Tansley, A. G. (1939) *The British Islands and their Vegetation*, Cambridge University Press, Cambridge.

Tauber, H. (1965) Differential pollen dispersion and the interpretation of pollen diagrams. *Geol. Surv. Denmark II Ser.*, **89**, 1–69.

Tilman, D. (1988) Plant strategies and the dynamics and structure of plant communities. *Princeton Monogr. Pop. Biol.*, **26**, 1–360.

Transeau, E. N. (1935) The Prairie Peninsula. *Ecology*, **16**, 423–37.

Tsukada, M. (1982a) *Pseudotsuga menziesii* (Mirb.) Granco: its pollen dispersal and late Quaternary history in the Pacific Northwest. *Jap. J. Ecol.*, **32**, 159–87.

Tsukada, M. (1982b) Late-Quaternary development of the *Fagus* forest in the Japanese archipelago. *Jap. J. Ecol.*, **32**, 113–8.

Urban, D. L., O'Neill, R. V. and Shugart, H. H. (1987) Landscape ecology: a hierarchical perspective can help scientists understand spatial patterns. *Bioscience*, **37**, 119–27.

Valentine, J. W. (1973) *Evolutionary Paleoecology of the Marine Biosphere*, Prentice-Hall, Englewood Cliffs, NJ.

Van der Pijl, L. (1982) *Principles of Dispersal in Higher Plants*, Springer-Verlag, New York.

Velichko, A. A., Wright, H. E., Jr. and Barnosky, C. W. (eds) (1984) *Late Quaternary Environments of the Soviet Union*, University of Minnesota Press, Minneapolis.

Von Post, L. (1916) Forest tree pollen in south Swedish peat bog deposits. *Pollen et Spores*, **9**, 378–401. (Translated by M. B. Davis and K. Faegri in 1967.)

Von Post, L. and Sernander, R. (1910) Pflanzen-physiognomische Studien auf Torfmooren in Narke. XI *International Geological Congress: Excursion Guide No. 14* (A7), Stockholm, 14 p.

Walker, D. (1970) Direction and rate in some British post-glacial hydroseres, in *Studies in the Vegetational History of the British Isles* (eds D. Walker and R. G. West), Cambridge University Press, Cambridge, pp. 117–39.

Walker, D. and Chen, Y. (1987) Palynological light on rainforest dynamics. *Quat. Sci. Rev.*, **6**, 77–92.

Watts, W. A. (1973) Rates of change and stability in vegetation in the perspective of long periods of time, in *Quaternary Plant Ecology* (eds H. J. B. Birks and R. G. West), Wiley and Sons, New York, pp. 195–206.

Watts, W. A. (1980a) The late Quaternary vegetation history of southeastern United States. *Ann. Rev. Ecol. Syst.*, **11**, 387–409.

Watts, W. A. (1980b) Late-Quaternary vegetation history at White Pond on the Inner Coastal Plain of South Carolina. *Quat. Res.*, **13**, 187–99.

Watts, W. A. (1988) Europe, in *Vegetation History* (eds B. Huntley and T. Webb, III), Kluwer, Dordrecht, pp. 155–92.

Webb, L. J. and Tracey, J. G. (1981) Australian rainforests: patterns and change, in *Ecological Biogeography of Australia*, Vol. 1 (ed. A Keast), Dr W. Junk, Dordrecht, pp. 605–94.

Webb, S. L. (1986) Potential role of passenger pigeons and other vertebrates in the rapid Holocene migrations of nut trees. *Quat. Res.*, **26**, 367–75.

Webb, S. L. (1987) Beech range extension and vegetation history: pollen stratigraphy of two Wisconsin lakes. *Ecology*, **68**, 1993–2005.

Webb, T. III (1980) The reconstruction of climatic sequences from botanical data. *J. Interdisciplinary Hist.*, **10**, 749–72.

Webb, T. III (1988) Eastern North America, in *Vegetation History* (eds B. Huntley and T. Webb III), Kluwer, Dordrecht, pp. 385–414.

Webb, T. III and Bryson, R. A. (1972) Late- and postglacial climatic change in the northern Midwest, USA: quantitative estimates derived from fossil pollen spectra by multivariate statistical analysis. *Quat. Res.*, **2**, 70–115.

Webb, T. III, Bartlein, P. J. and Kutzbach, J. E. (1987) Climatic change in eastern North America during the past 18 000 years; comparisons of pollen data with model results, in *North America and Adjacent Oceans During the Last Deglaciation*, Vol. K-3, *The Geology of North America* (eds W. F. Ruddiman and H. E. Wright, Jr.), Geological Society of America, Boulder, CO, pp. 447–62.

Webb, T. III, Cushing, E. J. and Wright, H. E., Jr. (1983) Holocene changes in the vegetation of the Midwest, in *Late-Quaternary Environments of the United States*, Vol. 2, *The Holocene* (ed. H. E. Wright, Jr.), University of Minnesota Press, Minneapolis, pp. 142–65.

West, D. C., Shugart, H. H. and Botkin, D. B. (eds) (1981) *Forest Succession, Concepts and Application*, Springer-Verlag, New York.

West, R. G. (1961) Interglacial and interstadial vegetation in England. *Proc. Linn. Soc. London*, **172**, 81–9.

West, R. G. (1977) *Pleistocene Geology and Biology*, Longman, London.

Wheeler, N. C. and Guries, R. P. (1982a) Biogeography of lodgepole pine. *Can. J. Bot.*, **60**, 1805–14.

Wheeler, N. C. and Guries, R. P. (1982b) Population structure, genetic diversity, and morphological variation in *Pinus contorta* Dougl. *Can. J. For. Res.*, **12**, 595–606.

White, J. (ed.) (1985) *Studies on Plant Demography*, Academic Press, London.

White, P. S. (1979) Pattern, process, and natural disturbance in vegetation. *Bot. Rev.*, **45**, 229–99.

Whitehead, D. R., Charles, D. F., Jackson, S. T., Reed, S. E. and Sheehan, M. C. (1986) Late-glacial and Holocene acidity changes in Adirondack (NY) lakes, in *Diatoms and Lake Acidity* (eds J. P. Smol, R. W. Battarbee, R. B. Davis and J. Merilainen), Dr W. Junk, Dordrecht, pp. 251–74.

Whiteside, M. C. (1983) The mythical concept of eutrophication. *Hydrobiologia*, **103**, 107–11.

Whitney, G. G. (1986) Relation of Michigan's presettlement pine forests to substrate and disturbance history. *Ecology*, **67**, 1548–59.

Whittaker, R. H. (1956) Vegetation of the Great Smoky Mountains. *Ecol. Monogr.*, **26**, 1–80.

Whittaker, R. H. (1975) *Communities and Ecosystems*, MacMillan, New York.

Winkler, M. G. (1988) Paleolimnology of a Cape Cod kettle pond: diatoms and reconstructed pH. *Ecol. Monogr.*, **58**, 197–214.

Winter, T. C. (1985) Physiographic setting and geologic origin of Mirror Lake, in *An Ecosystem Approach of Aquatic Ecology, Mirror Lake and its Environment* (ed. G. E. Likens), Springer-Verlag, New York, pp. 40–53.

Woods, K. D. and Davis, M. B. (1989) Paleoecology of range limits: beech in the Upper Peninsula of Michigan. *Ecology*, **70**, 681–96.

Wright, H. E., Jr. (1968) History of the Prairie Peninsula, in *The Quaternary of Illinois, Special Report 14* (ed. R. E. Bergstrom), College of Agriculture, University of Illinois, Urbana, pp. 78–88.

Wright, H. E., Jr (1976) The dynamic nature of Holocene vegetation, a problem in paleoclimatology, biogeography, and stratigraphic nomenclature. *Quat. Res.*, **6**, 581–96.

Wright, H. E., Jr. (1977) Quaternary vegetation history – some comparisons between Europe and America. *Ann. Rev. Earth Planet. Sciences*, **5**, 123–58.

Wright, H. E., Jr. (1980) Surge moraines of the Klutlan Glacier, Yukon Territory, Canada: origin, wastage, vegetation succession, lake development, and application to the late-glacial of Minnesota. *Quat. Res.*, **14**, 2–18.

Wright, H. E., Jr. (1981) Vegetation east of the Rocky Mountains 18 000 years ago. *Quat. Res.*, **15**, 113–25.

Wright, H. E., Jr. (ed.) (1983) *Late-Quaternary Environments of the United States*, Vol. 2, *The Holocene*, University of Minnesota Press, Minneapolis.

Wright, H. E., Jr. (1984) Sensitivity and response time of natural systems of climatic change in the late Quaternary. *Quat. Sci. Rev.*, **3**, 91–131.

Index